Advances in Applied Science and Engineering

nova
science publishers
www.novapublishers.com

Advances in Applied Science and Engineering

Research Challenges in Science, Engineering and Technology
M. Shanmuganantham, R. Nagarajan, B.E., M.E., PhD (Author),
S. Kannadhasan, PhD (Author)
2021. ISBN: 978-1-68507-008-3 (Hardcover)
2021. ISBN: 978-1-68507-028-1 (eBook)

COVID-19 Pandemic: Questions, Answers and Hypotheses
A. K. Haghi, PhD
2020. ISBN: 978-1-53618-565-2 (Softcover)
2020. ISBN: 978-1-53618-593-5 (eBook)

Innovations and Challenges in Natural Products Chemistry: Research and Applications
A. K. Haghi, PhD
2020. ISBN: 978-1-53618-650-5 (eBook)

Metamaterials for Microwave and Terahertz Applications: Absorbers, Sensors and Filters
Bhargav Appasani, PhD (Editor)
Om Prakash Acharya (Editor)
Amitkumar Vidyakant Jha (Editor)
Nisha Gupta (Editor)
2022. ISBN: 979-8-88697-330-3 (Hardcover)
2022. ISBN: 979-8-88697-419-5 (eBook)

More information about this series can be found at
https://novapublishers.com/product-category/series/advances-in-applied-science-and-engineering/

Vijay Kumar
Editor

Understanding Information Entropy

nova science publishers
www.novapublishers.com

https://doi.org/10.52305/GRXE4871

NOTICE TO THE READER

Library of Congress Cataloging-in-Publication Data

Names: Kumar, Vijay (Professor of mathematics), editor.
Title: Understanding information entropy / Vijay Kumar (editor)
Description: New York : Nova Science Publishers, [2023] | Series: Advances in applied science and engineering | Includes bibliographical references and index. |
Identifiers: LCCN 2023021936 (print) | LCCN 2023021937 (ebook) | ISBN 9798886978391 (hardcover) | ISBN 9798886978940 (adobe pdf)
Subjects: LCSH: Entropy (Information theory) | Measure theory. | Decision making--Mathematical models.
Classification: LCC Q370 .U53 2023 (print) | LCC Q370 (ebook) | DDC 003/.54--dc23/eng20230708
LC record available at https://lccn.loc.gov/2023021936
LC ebook record available at https://lccn.loc.gov/2023021937

Published by Nova Science Publishers, Inc. † New York

Contents

Preface

"We demand rigidly defined areas of doubt and uncertainty!"
—Douglas Adams, *The Hitchhiker's Guide to the Galaxy*

The actual concept of entropy was originated in 1948 with the landmark paper *A Mathematical Theory of Communication* by Claude Shannon (1948), which considered only statistical processes and was the average rate at which information being produced by a stochastic source of data. He defined the concepts of information in mathematical insight and proved many general results and their consequences. Since then, entropy has been extensively used in different domains including quantum information theory, communication theory, image recognition, finance, decision-making etc. Many generalizations of the entropies have been proposed that have applications in various domains of study for the purpose of better decision-making. The applications of information theoretic measures have been studied in this book with an aim to make better understanding of the use of entropy across domains. Also, generalized entropy measures have been studied that helps the theory to grow. Uncertainty arises due to the lack of some useful information, whereas, information measures help to measure and reduce the uncertainty in the system by the deficiency of information.

In general, information measures have been used to construct systems for various purposes such as: prediction, retrodiction, diagnosis, prescription, planning and control. In classical information theory, uncertainty is the primary concept and information is defined in terms of uncertainty reduction. In uncertainty theory, framework has been recognized as classical or fuzzy and the measure used for identification could be additive or non-additive, that allows the researchers to do research in uncertainty theories and develop some useful measures. Generalized information theory covers comprehensive results since its inception that aims to develop mathematical framework of the interrelated concepts of uncertainty and information in all their versions. Many proposed generalized information theoretic measures have been discussed in this book. The primary aim of the book is to provide the reader with a comprehensive and in-depth overview of the proposed generalized measures that have applications across domains. The details of these measure have been discussed in the following chapters.

Chapter 1 - The chapter focusses on various information theoretic measures in quantum systems, mainly prototypical systems such as double well (DW) potential both in symmetric as well as asymmetric, free as well as a confined hydrogen atom (CHA) enclosed in a spherical impenetrable cavity and a many-electron atom under similar enclosed environment. The generalized information theoretic measures such as Rényi (R) and Tsallis (T) are closely associated with entropic moments and are the generalization of Shannon Entropy (S). A pair of rules have been proposed for asymmetric double well (ADW) using energy results and information analysis and found that two wells of ADW behave as two different potentials. Also, suitability and applicability of the new virial theorem has been discussed.

Chapter 2 - The application of information theoretic measures in the description of molecular ensembles and discussed many interesting results of super-additivity and mixing of features of molecular ensembles have been demonstrated in this chapter. Also, result of magic ensembles characterized with zero information entropy of mixing has been proposed. These results build up the prospective of information-entropy-based discrete mathematical chemistry and related interdisciplinary studies.

Chapter 3 - A logarithmic entropy measure based on Shannon entropy for neutrosophic set has been proposed in this chapter. Also, a multi-criteria decision-making method based on the proposed single valued neutrosophic entropy has been established in which criteria values of the alternatives are single valued neutrosophic sets (SVNSs).

Chapter 4 - Intuitionistic fuzzy based TOPSIS method has been proposed, which is a multi- criteria decision-making approach. The application of the proposed method has been demonstrated with the help of case study of the selection of university for a particular student from the available pool of universities. This method considers multiple academic and non-academic factors and rank the university according to the features.

Chapter 5 - The generalized trigonometric information theoretic has been proposed, known as trigonometric intuitionistic fuzzy entropy measure. The measure has been validated through mathematical proof of the theorems. Also, the consistency of the proposed measure has been discussed with the help of numerical method.

Chapter 6 - The generalized information intuitionistic fuzzy parametric entropies measures of Rényi and Tsalli have been considered. The application of measures in the decision-making has been demonstrated by taking the case of Insurance sector.

Chapter 7 - The efficiency of stock market has been tested by evaluating the Volatility Index to assess the risk-neutral density. A modified entropy version of Black and Sholes' pricing model (1973) has been that study the observations on the call option of the Volatility Index (VIX) traded on the Chicago Board Option Exchange (CBOE), over the period starting from March 14th, 2018. From the study, it is concluded that the proposed method determine the arbitrary dynamics of equity, interest rates and volatility including implied volatility structures.

Acronyms

ADW	Asymmetric DW
CBOE	Chicago Board Option Exchange
CHA	Confined Hydrogen Atom
CHA	Confined H Atom
DFT	Density Functional Theory
DW	Double Well
FHA	Free Hydrogen Atom
IFE	Intuitionistic Fuzzy Entropy
IFS	Intuitionistic Fuzzy Set
IR	Relative Fisher Information
IvIFSs	Interval Valued Intuitionistic Fuzzy Sets
JPOEP	Multi-Configuration Parametrized Optimized Effective Potential
MCDM	Multi-Criteria-Decision-Making Approach
MED	Density of Maximum Entropy
QHO	Quantum Harmonic Oscillator
SDW	Symmetric Double Well
SE	Shannon Entropy
SVNS	Single Valued Neutrosophic Set
TOPSIS	Technique For Order Performance by Similarity to Ideal Solution

Chapter 1

Information Entropy in Excited States in Confined Quantum Systems

Sangita Majumdar
Neetik Mukherjee
and Amlan K. Roy*
Department of Chemical Sciences,
Indian Institute of Science Education and Research (IISER,
Kolkata, India

Abstract

The present contribution constitutes a brief account of information theoretical analysis in several representative model as well as real quantum mechanical systems. There has been an overwhelming interest to study such measures in various quantum systems, as evidenced by a vast amount of publications in the literature that has taken place in recent years. However, while such works are numerous in so-called *free* systems, there is a genuine lack of these in their constrained counterparts. With this in mind, this chapter will focus on some of the recent exciting progresses that has been witnessed in our laboratory [1, 2, 3, 4, 5, 6, 7, 8, 9, 10, 11, 12, 13, 14, 15, 16, 17], and elsewhere, with special emphasis on following prototypical systems, namely, (i) double well (DW) potential (symmetric and asymmetric) (ii) *free*, as well as a *confined hydrogen atom* (CHA) enclosed in a spherical impenetrable cavity (iii) a many-electron atom under similar enclosed environment.

*Corresponding Author's Email: akroy@iiserkol.ac.in, akroy6k@gmail.com

In: Understanding Information Entropy
Editor: Vijay Kumar
ISBN: 979-8-88697-839-1

The information generating functionals, Rényi (R) and Tsallis (T) entropies, are closely connected to entropic moments, completely characterizing density. Shannon entropy (S), Onicescu energy (E) are two particular cases of R, T; former signifying a measure of uncertainty, whereas latter quantifies separation of density with respect to equilibrium. Fisher information (I) is a gradient functional of density, identifying local fluctuation of a space variable. Some lower bounds (which do not depend on quantum numbers) are available for R, S; on the other hand, both upper and lower bounds of I have been achieved, which strictly vary with quantum numbers. Another related concept is Kullback-Leibler divergence or *relative entropy*, that delineates how a probability distribution changes from a given reference (usually chosen as ground state) distribution. In other words, this quantifies shift of information from one state to another.

A DW oscillator, $v(x) = \alpha x^m - \beta x^n + \gamma x^l$ (α, β, γ are real numbers, m, n, l are integers and $m > n > l$) may be classified into two categories (i) symmetric DW (SDW), where m, n are *even* and $\gamma = 0$ (ii) asymmetric DW (ADW), where m, n *even*, $\gamma \neq 0$ and l *odd*. They have their relevance in many physical and chemical phenomena, such hydrogen bonding, umbrella flipping of NH_3, anomalous optical lattice vibrations, proton transfer in DNA, internal rotation, etc. Through an accurate solution of Schrödinger equation by means of a variation-induced exact diagonalization procedure involving harmonic oscillator basis, these measures have been calculated. These, along with a phase-space analysis nicely explains the competing behavior of localization and delocalization. Appearance of *quasi-degeneracy* between lower even and adjacent upper odd states can be interpreted using these measures. In ADW a particle in a definite excited state can reside either in deeper or shallower well, depending upon the strength of the asymmetric parameter γ. In this context, a pair of rules have been proposed for ADW using energy results and the information analysis. It is found that at certain range of γ, two wells of ADW effectively behave as two different potentials.

Next we would move to the realistic atomic systems. It is interesting to note that the effect of confinement on excited quantum states is more dramatic. A detailed investigation of these information quantities along with several *complexity* (in both r and p spaces) would be probed for a caged-in atom. As it appears, some analytical results are possible, which would be considered accordingly. We will also discuss about the suitability and applicability of a newly designed virial theorem. For many-electron atoms, the relevant Kohn-Sham equation would be solved by adopting a work-function-based exchange potential, along with a suitable correlation energy functional, within the broad domain of *density functional theory*.

PACS: 03.65-w, 03.65Ca, 03.65Ta, 03.65.Ge, 03.67-a.

Keywords: Rényi entropy, Shannon entropy, Fisher information, Onicescu energy, complexities, confined harmonic oscillator, confined H atom, confined many-electron atom, excited state

1. Introduction

From the early days of quantum mechanics the problem of a *particle-in-a-box* has played the role of a simple, and particularly educative example, showing the difference in energy spectrum between a *free* and a *confined* quantum particle. By *free* system, we refer to quantum problems solved in whole space. Analysis of systems in some sub-region Ω of space is important mainly when one tries to model realistic situations in highly inhomogeneous media or in intense external fields. But during the last few years, it has become much more than a pedagogical tool and has found widespread applications in numerous areas of physics and chemistry. Matter under extreme conditions of pressure [18, 19, 20, 21] has been an interesting topic for a long time, for both experimental and theoretical researchers [22, 23, 24, 25, 26, 27, 28, 29, 30, 31, 32, 33, 34]. The experimental techniques have given the much needed insight about response of matter under such conditions. Because of a wide variety of applications, considerable attention has been given to analyze the effect of very high pressure on electronic structure, energy spectrum, chemical reactivity, ionization potential, molecular bond size, polarizabilities of atoms and molecules confined by cavities of different geometrical forms and dimension [23, 18, 19, 20, 35, 36, 37], etc. Apart from these, plenty of other applications are also reported, e.g., cell model in liquid sate, semiconductor dots, atoms encapsulated within nanocavities, such as fullerene, zeolite molecular sieves, solvent environments, porous silicon. Some other systems of interest are confined phonons, polaritons, plasmons, confined gas of bosons etc. These models of spatial confinements are also of significance in astrophysics for understanding of mass-radius relation of white dwarfs and ionized plasma properties. In order to describe states of atoms or molecules in cavities or in semiconductor nanostructures, one could formally perform extensive large-scale *first principles* calculations for clusters containing embedded objects. In such a case one might get a proper description of the whole system, but still it is difficult to routinely perform such elaborate calculations, and may not necessarily lead to simple interpretation of the results gained. While such attempts have grown more with time, still it is quite desirable to look for simpler methods to understand confinement effects.

In general, the effect of spatial confinement is of same origin as some other phenomena caused by electric fields in a cavity environment. But wherever it is possible, it is usually convenient to consider them separately, distinguishing the barrier-type sharp border potentials from other ones. Spatial confinement can thus, in the simplest way, be modeled by putting potential barriers at borders of a confining region. In such a scenario, the behavior of wave function at cavity border, and its outside, is forced by the presence of the wall. In some occasions, however, potential barriers do not appear directly in equations, and we have to impose boundary conditions that reflect the fact that, probability of finding the object outside its region of confinement is zero or nearly zero. This means requiring the vanishing (or almost) of the wave function at cavity border, which, on the other hand, corresponds to the presence of infinite/finite potential barriers. For a wide range of physical situations, one may consider the Schrödinger equation for a given subsystem and use an appropriate nontrivial boundary condition on the boundary $\delta\Omega$ of the concerned region Ω. In order to connect some of the basic concepts, we start by pointing out the difference between Dirichlet, Neumann and Robin types. In traditional mathematical physics, one considers the following self-adjoint boundary conditions on wave functions:

$$
\begin{aligned}
\psi(\mathbf{r})|_{\mathbf{r}\in\delta\Omega} &= 0 \quad \text{(Dirichlet boundary condition)}, \\
\delta_n\psi(\mathbf{r})|_{\mathbf{r}\in\delta\Omega} &= 0 \quad \text{(Neumann boundary condition)}, \\
\delta_n\psi(\mathbf{r}) - P(\mathbf{r})\psi(\mathbf{r})|_{\mathbf{r}\in\delta\Omega} &= 0 \quad \text{(General boundary condition)},
\end{aligned} \tag{1}
$$

where δ_n is the normal derivative, i.e., $\delta_n\psi = (\mathbf{n}, \nabla\psi)$, for a unit vector $\mathbf{n}$ externally normal to the surface $\delta\omega$ at a given point, and $\nabla\psi$ is the gradient of wave function ψ, while $P(\mathbf{r})$ denotes some real-valued function. The Neumann boundary condition implies $P \equiv 0$, while for Dirichlet boundaries one may suppose formally $|P| \rightarrow \infty$.

Probably one of the first papers that analyzed an atomic system in a bounded region, was due to Wigner and Seitz [38, 39] on the theory of periodic structures, published in 1933-1934. The Schrödinger equation for an atom in a lattice was studied with Neumann boundary conditions. In this work Ω corresponded to a sphere of radius R and the nucleus was placed at its center. This helped reduce the problem to that of radial function only. Later, Michels *et al.*, [23] considered a hydrogen atom confined within a sphere built with rigid walls, such that the electron density, $\rho(\mathbf{r})$, vanishes at the boundary of confining radius.

This publication is often attributed as first example of how Dirichlet boundary condition can be applied to study real physical problems. Note that the approach is still in use, especially in astrophysics. Prior to this work, however, in 1911, Weyl solved some vibrational problems, which may be interpreted as describing the structure of highly excited part of the spectrum of a particle in a bounded region Ω with Dirichlet boundary conditions.

Confinement model is very useful to examine effects on atoms trapped in a microscopic cavity. In order to investigate pressure and polarizabilty as a function of compression, a model consisting of a hydrogen atom at the centre of an impenetrable box, was proposed [23], which has become very popular in dealing with confined quantum systems with boxes of different size and geometrical forms. Similarly, many-electron atoms subjected to extreme pressure can also be simulated by placing them inside an impenetrable cavity of adjustable radius. Here, the infinite potential is induced by neighboring particles of negative charge. In literature, this kind of trapping is known as hard confinement. The electrostatic Hamiltonian is modified by adding a confining potential in term of radius r_c. However, this only includes effects produced by repulsive forces. To account for the existence of attractive forces between particles, such as van der Waals forces, it was proposed that the potential surface be finite. Confined atoms by penetrable confining potentials were also scrutinized by modifying adequately the extra potential added to the Hamiltonian. It may be emphasized that the incorporation of a particular boundary condition poses significant difficulties in calculation of the corresponding energy spectra. Hence the success of a given method in ground state may not necessarily be carried forward in excited states. Recent progresses in understanding of caged atoms taking into account various confining environments and their modeling, changes in their properties, as well as the methods of analysis and solutions along with their accuracy, are illustrated in the next section.

An immense amount of theoretical works have been published over several decades, covering a broad variety of confined systems. It has been analyzed by putting the atom inside boxes of different geometrical forms (with varying size within hard boundary); the most preferred choice is a spherical box of penetrable or impenetrable walls. A confined H atom (CHA) within an impenetrable [40, 41, 29, 42, 43, 44, 45, 46, 47, 4, 5] (as well as penetrable) cavity was investigated quite vigorously, as recognized to be an original treasure-trove of numerous attractive properties, both from physical and mathematical point of view. Finding approximate analytical as well as numerical solutions, and comparing their accuracy to exact solution has been an active field of research. A broad range of theoretical methods varying in difficulty, sophistication and accuracy is available; a selected set includes perturbation theory, Padé approximation, WKB method, Hypervirial theorem, power-series solution, super-symmetric quantum mechanics, Lie algebra, Lagrange-mesh method, asymptotic iteration method, generalized pseudospectral method [40, 41, 29, 42, 43, 44, 45, 46, 47, 4, 5] etc., and the references therein. Exact solution [43] of CHA has been put forth in terms of Kummer M-function (confluent hypergeometric). While CHA energy spectrum shows a monotonic, unlimited increase of its energies as the boundary approaches nucleus and volume of confinement is reduced, the same in case of open conoidal boundaries (sphere, circular cone, paraboloid, prolate spheroid, hyperboloid) is characterized by a monotonic increase of energy levels only up to zero energy in the corresponding limit situations, with consequent infinite degeneracy. The exact solution of Schrödinger equation for hydrogen atom in spherical, spheroconal, parabolic and prolate spheroidal coordinate has also been reported by several researchers. In contrast to the free hydrogen atom (FHA), in CHA due to symmetry breaking, different energy eigenvalues, eigenfunctions and reduced degeneracies is observed. Study of CHA with soft spherical boxes can also be found in literature where the compression regimes are considered with Neumann boundary conditions. They offer many unique phenomena, especially relating to *simultaneous, incidental and inter-dimensional degeneracy* [48, 15]. Effect of compression on ground and excited energy levels, as well as other properties, like hyper-fine splitting constant, dipole shielding factor, nuclear magnetic screening constant, pressure, static, and dynamic polarizability, were probed.

We now extend our discussion to many-electron atoms confined in various environments. The analysis of such systems are complicated due to the electron-electron repulsion term. Unlike hydrogen, in such systems, the electron correlation, which is an essential component of electronic structure, can be studied. As confinement by hard impenetrable boxes overestimates the response of physical observables, penetrable walls are more-convenient to mimic the experimental counterpart. Confined atoms centered within a sphere enveloped by *hard walls*–a widely used model to imitate enclosed environment, can be considered as a starting point to the understanding of confinement inside *penetrable walls* [37, 19, 36, 49, 50, 51, 52]. Helium atom being the prototypical many-electron atom and one of the most abundant elements found to exist as a significant component of giant planets and stars, is a most relevant one to approach the electronic structure and spectroscopic properties, when subjected to high pressure. Calculating electronic properties of these systems, in which electronic clouds are forced to remain spatially restricted, represents a more demanding task. The investigations on He confinement was carried out fifteen years after the study of hydrogen atom by Seldam *et al.* [53] where they performed a variational calculation with Hylleraas wave function, multiplied by a cut-off function, for ground state. Thereafter, several other analyses have been extended to determine the electronic structure of ground as well as few low lying excited states, which can be broadly categorized as: i) wave-function based and ii) density functional theory (DFT) techniques. The Hartree-Fock (HF) and Kohn-Sham (KS) models are the most popular and promising routes to obtain wave function and electron density, although variational approaches based on the generalized Hylleraas basis set (where the wave function contains a cut-off factor to ensure Dirichlet boundary condition) has also been employed towards their understanding [54, 55, 56, 57, 58]. There are basically two ways to obtain the wave function or electron density under these boundary conditions: using a cut-off function multiplied with basis functions in restricted HF framework to impose such a confinement [32, 59, 60] or imposing the behavior on wave function within the numerical scheme when solving SCF method for the radial equation [35, 42]. Another alternative variational method is the so-called optimized effective potential approximation, which has been generalized to work with multi-configuration wave function. This was applied to compute electronic structure of both ground and excited states in constrained atoms [61, 62, 63]. Several other noteworthy techniques that were engaged towards these are: Rayleigh-Ritz [64, 28], linear variational [65], HF [66, 35], configuration interaction [67], quantum Monte Carlo [68], etc. Here it is worth mentioning that, although there exist several wave-function based methods for these kind of systems, density-based attempts are rather very scarce [42, 54]. Apart from these, considering the nucleus to be fixed at the centre of spherical box, Whitkop [69] achieved a novel contribution upon studying with nucleus placed off centre of the impenetrable spherical box.

In the last few decades, the concept of quantum information theory has emerged as a subject of topical interest due to its extensive applications [70, 71, 72] in understanding various phenomena in physics and chemistry, relating to diverse topics such as thermodynamics, quantum mechanics, spectroscopy etc. In recent years, it has also been used prodigiously in the study of quantum entanglement and quantum steering problems. In literature, the information-theoretic measures like Shannon entropy (S) [73, 74], Rényi entropy (R^{α}) [75], Fisher information (I), Onicescu energy (E) and complexities (C) are invoked to explore multitude of phenomena such as, diffusion of atomic orbitals, spread of electron density, periodic properties, correlation energy and so forth. Arguably, entropic uncertainty relations, constructed from these fundamental information theoretical tools, are the most effective measures of uncertainty [73, 75, 76, 77], as they do not make any reference to some specific points of the respective Hilbert space. In a quantum system, S and I quantify the information content in a complimentary way. Former is a global measure of spread of density which refers to the expectation value of logarithmic probability density function. On the other hand, I is a gradient functional of density and it signifies the oscillatory nature and sharpness of density in position space (r). It is well known that, R^{α}, T^{α}, the so-called information generating functionals, are closely connected to entropic moments (discussed later), and completely characterize density $\rho(\mathbf{r})$. They actually quantify the spatial delocalization of single-particle density of a system in various complimentary ways. Moreover, these are closely related to energetic and experimentally measurable quantities [78, 79]. It is interesting to note that, S, E (disequilibrium) are two particular cases of R^{α}, T^{α} [79, 75]. Former measures total extent of density whereas E quantifies separation of density with respect to equilibrium. However, $(-)$ve value in R and S indicates extreme localization, whereas, E is always $(+)$ve. Likewise, changing the numerical values of R^{α}, S from $(-)$ve to $(+)$ve only interprets enhancement of de-localization. Another associated concept is *complexity*. A system has finite complexity when it is either in a state with less than some maximal order, or not at a state of equilibrium. In a nutshell, it becomes *zero* at two limiting cases, *viz.,* when a system is (i) completely ordered (maximum distance from equilibrium) or (ii) at equilibrium (maximum disorder). Complexity has its contemporary interest in chaotic systems, spatial patterns, language, multi-electronic systems, molecular or DNA analysis, social science, astrophysics and cosmology etc [79].

A particle in an SDW potential is a prototype of a system for which perhaps, a full quantum mechanical explanation of confinement is more effective than that of traditional uncertainty product relation. This has relevance in unraveling a wide variety of physical, chemical phenomena of a model two-state system [80], wherein one may identify two wells as two different states of a quantum particle. Some important applications worth recording are, hydrogen bonding, umbrella flipping of ammonia molecule, anomalous optical lattice vibrations in

HgTe, proton transfer in DNA to model brain micro-tubules, internal rotation, physics of solid-state devices, solar cells and electron tunneling microscopes, etc. Existence of an interplay between localization and delocalization effects is another interesting feature of this system. These competing effects lead to a number of quasi-degenerate pair states. An increase in positive term (strength) reduces spacing between classical turning points but at the same time, reduces barrier area as well as barrier height. Conversely, negative term increases spacing between classical turning points but also increases barrier area and barrier height. It has been pointed out that the characteristic feature of quasi-degeneracy present in a DW potential can not be examined from commonly used uncertainty relation.

Introduction of an asymmetric term in the DW potential makes it more anomalous and interesting. Because it gives rise to two asymmetric wells. From a classical mechanical point of view, a particle will always reside in deeper well, but in quantum mechanics the situation is not that straightforward. Some typical applications of asymmetric DW potential are: electron in a double quantum dot, Bose-Einstein condensation in trapped potentials, quantum super-conducting circuit based on Josephson junction, quantum computing devices, etc. The contrasting effects present in such systems, especially in excited states, can be beautifully analyzed in terms of information entropic measures.

Next, in this paragraph, we shift towards the information in atoms. In recent years, appreciable attention has been paid to investigate various information measures in central potentials. At first we restrict ourselves to a few references pertaining to H atom–both FHA and CHA. Some of these for FHA are: I_r, I_p, I in 3D [81], in D-dimension [82, 83, 84]; upper bounds of S, R [85], S in 3D [86], in D-dimension [87]; R, T in 3D [86], in D-dimension [88]. Relativistic effects on the information measures of FHA are also examined [89]. A lucid review on information theory of D-dimensional FHA is provided in [90]. The exact mathematical form of I for an arbitrary state of FHA was given in both r and p space [81] in terms of four expectation values $\langle r^2 \rangle$, $\langle p^{-2} \rangle$, $\langle p^2 \rangle$, $\langle p^{-2} \rangle$, and eventually can be expressed in terms of related quantum numbers. Likewise, an exact analytical formula for S in ground state of a D-dimensional FHA was derived long times ago [87] in both r and p space. Later, similar analytical expressions of S for circular or node-less states of a D-dimensional FHA was offered in 2010 [90] in both spaces. However, such a closed form expression of S is as yet lacking for a general state. Very recently, in [91], accurate radial Shannon entropies in r, p spaces, with $n = 10$ are computed numerically.

Moreover, a generalized form of angular Shannon entropy was also derived. Recently, R and T for Rydberg hydrogenic states were reported within a strong Laguerre asymptotic approximation. [86, 88]. But their exact closed-shell forms are as yet unknown; moreover these were produced so far, only in r space, and mostly for $l = 0$ states. For FHA, all these can be calculated from exact analytical wave functions in r, p spaces. Recently we have found that, for node-less states, expressions of S, R, T, E are accessible in closed form in a FHA, as the required radial polynomial reduces to unity. But for all other n, l they need to be computed numerically, as the presence of nodes in such wave functions leads to difficult polynomials.

However, analogous studies as above, in CHA are quite scanty and are of recent origin; e.g., S [48, 91, 11], bounds of I, S, R, T [92], as well as I, S in case of soft spherically CHA. [93]. Apart from the work [91] for S, it was analyzed in the context of soft and hard confinement in lowest state [48, 93]. The variation of S in s, p, d orbitals ($n \leq 7$) was followed [91] with r_c. A detailed systematic variation of these measures in CHA, with respect to r_c (as well as FHA) has been presented only lately [11, 13]; for non-zero angular momentum they are separately published in [94]. hydrogen−like ions in conjugate r and p spaces has been done in our lab. [13]. Besides these independent information quantities, several well-known measures similar to LMC and Fisher-Shannon complexity have been pursued for CHA in both conjugate spaces [8].

When one analyzes these aforementioned quantities for many-electron systems, electron correlation comes into play. By treating the electron probability distribution as a continuous function, these quantities are applied in the study of different model potentials mimicking diatomic molecular potentials [95] and numerous atomic and molecular systems [96, 97, 98, 99, 100, 101, 102, 88, 86, 103]. Amongst these, S is the hallmark of information theory [104, 105]. From the standpoint of not being able to obtain any analytic forms, significant amount of numerical study has been published for innumerable many-body systems. For instance, an entropy maximization technique was applied to explain the Compton profiles of atoms [106]. Later, Gadre *et al.* [107] considered S within a Thomas-Fermi model. They also speculated that Shannon entropy sum, (S_t) can be used as an indicator of quality of N-electron wave functions [108]. This was extended for similar works [109] in arbitrary D-dimensional many-particle system. A validation came from an investigation [110] of electronic correlation in Be-isoelectronic series. A numerical HF calculation [79] for free neutral atoms suggested that S_t could play the role of an indicator of shell structure.

Beyond HF-level calculations were reported as well [111, 112]. S has various other applications as well in the study of free atoms, namely, descriptor of several properties of a finite coulomb system in both ground and excited states [104], measure of aromaticity [113], manifests in avoided crossing [78], quantifies amount of electronic interactions present in a system [114], measure of correlation [96] and so on. I has been demonstrated to be a very useful tool in analyzing atoms [81, 115, 116, 117, 118, 119, 120, 121, 122, 123, 124]. Moreover, S, I have been depicted as an indicator of quality of an approximate wave function [116]. Though a large-scale of study for these quantities are available in literature for free atoms, amount of similar study for confined systems is very limited. Sen [48] has examined S for H-and He-like systems with impenetrable boundary. The author tried to locate if there exists any optimum point in S_t as a function of confinement radius. Ou *et al.* [125] pursued similar analysis for He confined with penetrable potential in its ground and excited states by employing highly correlated and extensive Hylleraas-type wave function. HF study of electron-delocalization in other many-electron atoms confined by penetrable walls in terms of S has also been executed [126]. The present authors have investigated these information theoretic measures for atoms and ions under high pressures in both ground and excited states employing a work function based strategy within the framework of DFT [14, 16, 17].

Besides the aforementioned information tools, complexity measures also have the ability to estimate a variety of physical or biological systems for organization. Statistical measure of complexity were presented for atoms using HF densities in both r and $p-$ spaces where analysis reveals interesting correlations with quantities such as ionization potential, static dipole polarizability, shell structure of electron density etc. [20]. Apart from one-electron densities, two-electron atomic densities (pair density) were also studied in terms of complexity measures (shape complexity) and its corresponding information planes [127]. In addition to atomic systems, it has applications in diverse fields like detection of periodic, quasi-periodic, linear stochastic, chaotic dynamics etc.

The present contribution gives an outline of the recent progresses that have taken place in the information theoretical analysis of quantities, like S, R, I, E, C, in the ground and excited states of several quantum confined systems, such as a particle confined in SDW and ADW potential, CHA. Furthermore, we have also dealt with a KS model, which has been applied successfully for calculation of energy spectrum of many-electron atoms under similar kind of confinement. In the 1D DW potential, we introduce a variation-induced exact diagonalization method for the accurate solution of relevant eigenvalue equation developed in our laboratory. In the 3D counterpart for confined atomic systems, we employ the very accurate generalized pseudospectral (GPS) method. These solutions are used throughout the chapter for the calculation of desired quantities presented. Detailed results are compiled in tabular and graphical form. Their changes are carefully monitored as one goes from unbounded to bounded system. Additionally we also offer a brief account of the relative information studies that has been undertaken by us recently. Available literature results have been consulted as and when possible. A few concluding remarks are made at the end.

2. Theoretical Aspects

This section provides the required details for theoretical solution of relevant eigenvalue equations as well as calculation of information-related quantities, as probed in this chapter. To simplify the discussion, three categories are made. At first, in order to achieve accurate eigenvalues and eigenfunctions in 1D DW potentials, a variation induced exact diagonalization method, involving manifold energy minimization scheme is adopted in both x and p spaces. In the second class, we deal with CHA, whole exact wave functions are available in the form of Kummer confluent hypergeometric function; respective energies are obtained via accurate GPS method. Finally, in the context of multi-electron atoms, we present a work function-based strategy; here also we engage the GPS method for solution of KS equation.

2.1. Variation-induced Exact Diagonalization

It is a simple basis-set method, and in principle, exact energies can be achieved provided the basis is complete. For convenience, the Hermite basis is adopted. Then the Hamiltonian can be represented in terms of raising and lowering operator of a conventional quantum harmonic oscillator (QHO), characterized by a potential of the form $V(x) = 4\sigma^2 x^2$. Thus, a QHO number operator basis with a single non-linear parameter σ (σ is related to the force constant k as $\sigma^2 = \frac{1}{8}k$) was utilized. In all the calculations that follow, dimension of the Hamiltonian matrix is set at $N = 100$ to ensure convergence of eigenvalues. Employing such a near-complete basis, matrix elements of Hamiltonian are defined as,

$$h_{mn} = \langle m|\hat{H}|n\rangle. \tag{2}$$

Diagonalization of the matrix h provides the energy eigenvalues and corresponding eigenvectors, which is easily performed by MATHEMATICA package.

In this work, instead of minimizing the individual eigenvalue, a manifold-energy minimization method [128, 129] was invoked. In this scheme, the trace of the matrix with respect to σ is minimized, which remains invariant under diagonalization of the matrix. Hence, one needs to diagonalize the matrix at that particular σ value for which trace of the matrix is minimum. This leads to all desired eigenvalues in a single diagonalization step.

The eigenvectors in position space then can be written in following form,

$$\psi_n(x) = \sum_{m=0} b_m^n \left(\frac{2\sigma}{\pi}\right)^{\frac{1}{4}} \frac{1}{\sqrt{2^m m!}} H_m(\sqrt{2\sigma}x)\, e^{-\sigma x^2} \tag{3}$$

while in momentum space, they are manifested as,

$$\psi_n(p) = \sum_{m=0} c_m^n \left(\frac{1}{2\sigma\pi}\right)^{\frac{1}{4}} \frac{1}{\sqrt{2^m m!}} H_m\left(\frac{p}{\sqrt{2\sigma}}\right) e^{-\frac{p^2}{4\sigma}}. \tag{4}$$

2.1.1. Symmetric Double-well (SDW) Potential

For our purposes, the SDW potential is conveniently written as,

$$V(x) = \alpha x^4 - \beta x^2 + \frac{\beta^2}{4\alpha}. \tag{5}$$

It is symmetric around $x = 0$ and two minima are located at $x_0 = \pm\sqrt{\frac{\beta}{2\alpha}}$, while the maximum value of potential is $\frac{\beta^2}{4\alpha}$ [130]. An increase in quartic parameter, α shortens separation between classical turning points and reduces barrier strength. Thus, on one side, it promotes localization and on other hand, minimizes confinement on the particle within a well. A rise in β causes duelling effects on particle. It leads to an enhancement in separation of classical turning points implying added delocalization of particle, whereas at the same time, barrier area and barrier height increase, promoting localization into one of the wells.

The Hamiltonian in position space is expressed as:

$$\hat{H}_x = -\frac{d^2}{dx^2} + \alpha x^4 - \beta x^2 + \frac{\beta^2}{4\alpha}, \tag{6}$$

whereas in momentum space, this reads as below,

$$\hat{H}_p = p^2 + \alpha\frac{d^4}{dp^4} + \beta\frac{d^2}{dp^2} + \frac{\beta^2}{4\alpha}. \tag{7}$$

For sake of convenience, we choose $m = \frac{1}{2}$ and $\hbar = 1$.

In position space, the non-zero matrix elements are expressed as [6],

$$\begin{aligned}
h_{lm} &= \frac{\alpha}{16\sigma^2}\sqrt{l(l-1)(l-2)(l-3)}, && \text{if } l-m=4 \\
&= \left[\frac{\alpha}{8\sigma^2}(2l-1) - \frac{(\beta+4\sigma^2)}{4\sigma}\right]\sqrt{l(l-1)}, && \text{if } l-m=2 \\
&= \left[2\sigma(2l+1) - (\beta+4\sigma^2)\frac{(2l+1)}{4\sigma} + \frac{3\alpha}{16\sigma^2}(2l^2+2l+1)\right], && \text{if } l-m=0 \\
&= \left[\frac{\alpha}{8\sigma^2}(2l-1) - \frac{(\beta+4\sigma^2)}{4\sigma}\right]\sqrt{(l+1)(l+2)}, && \text{if } l-m=-2 \\
&= \frac{\alpha}{16\sigma^2}\sqrt{(l+1)(l+2)(l+3)(l+4)}. && \text{if } l-m=-4
\end{aligned} \tag{8}$$

Conversely, in p space the non-zero terms are expressed in the following forms [6],

$$\begin{aligned}
g_{lm} &= h_{lm}, && \text{if } (l-m) = 4, -4 \\
&= -h_{lm}, && \text{if } (l-m) = 2, -2 \\
&= h_{lm}, && \text{if } (l-m) = 4, -4
\end{aligned} \tag{9}$$

The trace of h/g matrix in either spaces at a particular σ is same, given by,

$$Tr[h/g] = \sum_m h_{mm} =$$

$$= \sum_m \left[\frac{3\alpha}{16\sigma^2}(2m^2+2m+1) - \frac{(\beta+4\sigma^2)(2m+1)}{4\sigma} + 2\sigma(2m+1)\right] \tag{10}$$

Now, minimization of the trace leads to a cubic equation in σ,

$$8\sigma^3 + 2\sigma\beta - \alpha\,\frac{(2N^2 + 4N + 3)}{(N + 1)} = 0. \tag{11}$$

Use of parity, transforms Eq. (11), for even m as,

$$8\sigma^3 + 2\sigma\beta - \alpha(2N + 1) = 0, \tag{12}$$

while for odd m this offers,

$$8\sigma^3 + 2\sigma\beta - \alpha(2N + 3) = 0. \tag{13}$$

It can be easily demonstrated that σ will have a single real root in Eqs. (11), (12) and (13), for which the trace will be minimum for respective cases. Now, one can diagonalize the matrix employing the value of non-linear parameter σ.

2.1.2. Asymmetric Double Well (ADW) Potential

Without any loss of generality, the potential is simply written as [7],

$$V(x) = \alpha x^4 - \beta x^2 + \gamma x + V_0, \tag{14}$$
$$V_1(x) = \alpha x^4 - \beta x^2 - \gamma x + V_0. \tag{15}$$

Here V_0 signifies the values of global minimum of the potential. It has been added to make total energy positive. Clearly, V and V_1 constitute a mirror-image pair; substitution of x by $-x$ transforms one to other. These two potentials are *isospectral* in nature. Interestingly, their wave functions are also mirror images to each other. Thus, it is adequate to study the behavior of any one of them; other one automatically follows. This prompts us to choose $V(x)$ as our model ADW potential. Incorporation of a linear term in a SDW brings asymmetry in the system. Note that, α and β exert analogous effect to what we have observed in SDW potential. Introduction of asymmetric term shifts the maximum of potential from zero to either right or left sides depending on whether γ is positive or negative. Apparently, there exists a deeper (I) and a shallow (II) well. Figure 1 illustrates two typical scenario at two different γ values. It also portrays that maximum of V shifts towards right as γ increases.

The Hamiltonian in position space (for simplicity, we choose $2m = 1, \hbar = 1$) is:

$$\hat{H}_x = -\frac{d^2}{dx^2} + \alpha x^4 - \beta x^2 + \gamma x, \tag{16}$$

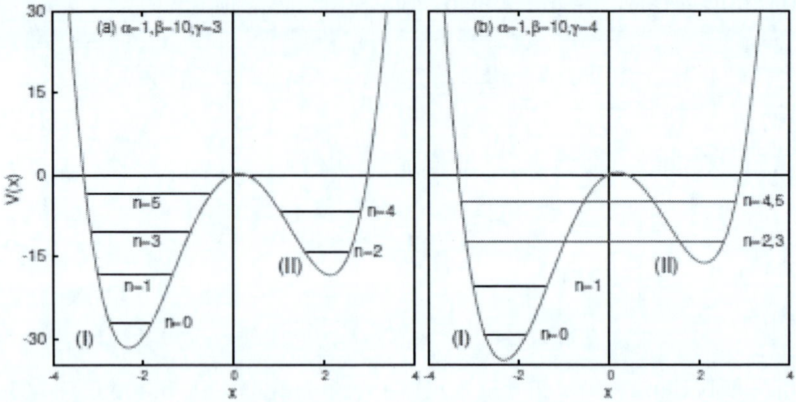

Figure 1. Schematic representation of an ADW potential, in Eq. (14), plotted at two different parameter sets: (a) $\alpha=1, \beta=10, \gamma=3$ (b) $\alpha=1, \beta=10, \gamma=4$. (I), (II) signify larger, smaller wells respectively. See text for details.

whereas in momentum space, this reads,

$$\hat{H}_p = p^2 + \alpha \frac{d^4}{dp^4} + \beta \frac{d^2}{dp^2} - i\gamma \frac{d}{dp}. \tag{17}$$

In position space, the non-zero matrix elements can be written as follows [7],

$$\begin{aligned}
h_{lm} &= \frac{\alpha}{16\sigma^2}\sqrt{l(l-1)(l-2)(l-3)}, && \text{if } l-m=4 \\
&= \frac{1}{8\sigma^2}[\alpha(2l-1) - 2(\beta+4\sigma^2)\sigma]\sqrt{l(l-1)}, && \text{if } l-m=2 \\
&= \gamma\sqrt{\frac{l}{4\sigma}}, && \text{if } l-m=1 \\
&= \frac{3\alpha}{16\sigma^2}(2l^2+2l+1) - (\beta+4\sigma^2)\frac{(2l+1)}{4\sigma} + 2\sigma(2l+1), && \text{if } l-m=0 \\
&= \gamma\sqrt{\frac{(l+1)}{4\sigma}}, && \text{if } l-m=-1 \\
&= \frac{1}{8\sigma^2}[\alpha(2l+3) - 2(\beta+4\sigma^2)\sigma]\sqrt{(l+1)(l+2)}, && \text{if } l-m=-2 \\
&= \frac{\alpha}{16\sigma^2}\sqrt{(l+1)(l+2)(l+3)(l+4)}. && \text{if } l-m=-4
\end{aligned} \tag{18}$$

On the other hand, in momentum space, the non-zero elements are obtained as

[7],

$$
\begin{aligned}
g_{lm} &= h_{lm}, && \text{if } (l-m) = 4, -4 \\
&= -h_{lm}, && \text{if } (l-m) = 2, -2 \\
&= -ih_{lm}, && \text{if } (l-m) = 1 \\
&= ih_{lm}. && \text{if } (l-m) = -1 \\
&= h_{lm}, && \text{if } (l-m) = 0 \qquad (19)
\end{aligned}
$$

Diagonalization of the *symmetric* matrix, **h/g** was performed efficiently by MATHEMATICA, leading to accurate energy eigenvalues and corresponding eigenvectors. We adopt a Manifold-Energy minimization approach [128, 129], where instead of minimizing a particular energy state, one minimizes trace of the matrix, which is given below,

$$
Tr[h] = \sum_l h_{ll} = \sum_l \left[\frac{3\alpha}{16\sigma^2}(2l^2 + 2l + 1) - \frac{(\beta + 4\sigma^2)(2l+1)}{4\sigma} + 2\sigma(2l+1) \right], \qquad (20)
$$

with respect to σ. This leads to a cubic equation in σ having only one real root. Finally the process is accomplished by minimizing the desired matrix in Eq. (20), for above value of σ.

2.2. Confined Atomic System

In this subsection, at first we discuss CHA and then a many-electron atom. The desired confinement is achieved by shifting the radial boundary from *infinity* to a *finite* range.

2.2.1. Position-space Wave Function

CHA: The exact radial wave function for a CHA can be expressed [44] as,

$$
\psi_{n,l}(r) = N_{n,l} \left(2r\sqrt{-2\mathcal{E}_{n,l}}\right)^l \times \qquad (21)
$$

$$
{}_1F_1 \left[\left(l + 1 - \frac{1}{\sqrt{-2\mathcal{E}_{n,l}}} \right), (2l+2), 2r\sqrt{-2\mathcal{E}_{n,l}} \right] e^{-r\sqrt{-2\mathcal{E}_{n,l}}},
$$

where $N_{n,l}$ indicates normalization constant and $\mathcal{E}_{n,l}$ is the energy of a given state characterized by n, l quantum numbers, whereas ${}_1F_1\,[a, b, r]$ illustrates confluent hypergeometric function. In case of FHA ($r_c \rightarrow \infty$), the first-order hypergeometric function modifies to associated Laguerre polynomial with $\mathcal{E}_{n,l} = -\frac{Z^2}{2n^2}$ (Z denotes atomic number). So the radial function simplifies to commonly used form, as given below,

$$\psi_{n,l}(r) = \frac{2}{n^2}\left[\frac{(n-l-1)!}{(n+l)!}\right]^{\frac{1}{2}} \left[\frac{2Z}{n}r\right]^l e^{-\frac{Z}{n}r}\, L^{(2l+1)}_{(n-l-1)}\left(\frac{2Z}{n}r\right). \tag{22}$$

Thus allowed energies at a specific r_c can be obtained by finding the zeros of ${}_1F_1$,

$${}_1F_1\left[\left(l+1-\frac{1}{\sqrt{-2\mathcal{E}_{n,l}}}\right), (2l+2), 2r_c\sqrt{-2\mathcal{E}_{n,l}}\right] = 0. \tag{23}$$

At a certain l, the first root corresponds to energy of the lowest-n state ($n_{lowest} = l+1$) with gradual roots identifying higher excited states. It is worthwhile noting that, in order to construct the exact wave function of CHA for a specific state, one requires to supply energy eigenvalue of that state. In our calculation, $\mathcal{E}_{n,l}$ of CHA, computed by means of the GPS method. The most important feature is that through an optimal, non-uniform spatial grid, it offers a symmetric eigenvalue problem, which can be easily solved by standard diagonalization routines, offering accurate eigenvalues and eigenvectors. Over the years, it has been applied to a variety of model and real systems, in both free and confined quantum systems, viz., spiked harmonic oscillators [131, 132], power-law and logarithmic [133], Húlthen and Yukawa [134], rational [135], Hellmann [136], exponential-screened Coulomb [137]; various molecular potentials for ro-vibrational states such as Morse [138], hyperbolic [139], shifted Deng-Fan [140], Tietz-Hua [141], Manning-Rosen [3], as well as other singular [142] potentials. Very successful applications have also been made in the context of many-electron systems within the broad domain of DFT [143, 144, 145, 146, 147, 148]. In past several years, this has also produced excellent quality results in various radial confinement studies in 3DQHO [1, 2, 4, 5].

At this moment, it is necessary to point out that, there exists an incidental degeneracy in CHA under following condition: **For all $n \geq (l+2)$, a (n, l) state is degenerate with $((n+1), (n+2))$ state at $r_c = (l+1)(l+2)$.** For example, at $r_c = 2$, energies of $1s$ and $3d$ states of CHA are degenerate; so are $2p$ and $4f$ states at $r_c = 6$

Many-electron atom: The starting point is the single-particle time-independent Kohn-Sham (TIKS) equation, which can be conveniently written as (atomic units employed, unless mentioned otherwise),

$$\mathbf{H}_0(\mathbf{r})\psi_i(\mathbf{r}) = \epsilon_i(\mathbf{r})\psi_i(\mathbf{r}), \tag{24}$$

where $\mathbf{H}_0$ is the unperturbed-KS Hamiltonian, given by

$$\mathbf{H}_0(\mathbf{r}) = -\frac{1}{2}\nabla^2 + \mathbf{V}_{eff}(\mathbf{r})$$
$$\mathbf{V}_{eff}(\mathbf{r}) = v_{ne}(\mathbf{r}) + \int \frac{\rho(\mathbf{r}')}{|\mathbf{r}-\mathbf{r}'|}\mathrm{d}\mathbf{r}' + \frac{\partial E_{xc}[\rho(\mathbf{r})]}{\partial\rho(\mathbf{r})}. \tag{25}$$

Though DFT has achieved impressive success in explaining electronic structure and properties of many-electron system in their ground state, calculation of excited-state energies and densities has remained a bottleneck. In this work, we employ a work-function method [149], according to which, exchange energy can be interpreted as interaction energy between an electron at $\mathbf{r}$ and its Fermi-Coulomb hole charge density $\rho_x(\mathbf{r}, \mathbf{r}')$, $\mathbf{r}$ at $\mathbf{r}'$,

$$E_x[\rho(\mathbf{r})] = \frac{1}{2}\int\int \frac{\rho(\mathbf{r})\rho_x(\mathbf{r},\mathbf{r}')}{|\mathbf{r}-\mathbf{r}'|}\,\mathrm{d}\mathbf{r}\mathrm{d}\mathbf{r}' \tag{26}$$

The exchange potential can then be defined as work done in bringing an electron to the point $\mathbf{r}$, against an electric field produced by its Fermi-Coulomb hole density.

$$\epsilon_x(\mathbf{r}) = \int \frac{\rho(\mathbf{r})\rho_x(\mathbf{r},\mathbf{r}')(\mathbf{r}-\mathbf{r}')}{|\mathbf{r}-\mathbf{r}'|^3} \tag{27}$$

The above expression gives the electric field due to Fermi-Coulomb hole charge density and from that the exchange potential can be obtained as,

$$v_x(\mathbf{r}) = -\int_{\infty}^{r} \epsilon_x(\mathbf{r})\mathrm{d}l \tag{28}$$

While the exchange potential $v_x(\mathbf{r})$ can be accurately calculated, the correlation potential $v_c(\mathbf{r})$ is unknown as yet, and must be approximated. We have taken into consideration two correlation energy functionals such as local Wigner [150] and non-local Lee-Yang-Parr (LYP) [151] functional. With this choice, this equation is solved numerically with GPS method. Now, to study the effect of hard confinement, we have imposed Dirichlet's boundary condition. The solution of TIKS equation obtained gives orthonormal atomic orbitals, from which the one-particle density can be found as,

$$\rho(\mathbf{r}) = \sum_{\mathrm{1}=1}^{N} |\psi_i(\mathbf{r})|^2. \tag{29}$$

Now information measures can be calculated in conjugate spaces, by standard procedure.

2.2.2. Momentum-space Wave Function

The p-space wave function ($\mathbf{p} = \{p, \Omega\}$) of a particle in a central potential is achieved from respective Fourier transform of its r-space counterpart [11], and is given below,

$$\begin{aligned}
\psi_{n,l}(p) &= \frac{1}{(2\pi)^{\frac{3}{2}}} \int_0^\infty \int_0^\pi \int_0^{2\pi} \psi_{n,l}(r)\ \Theta(\theta)\Phi(\phi)\ e^{ipr\cos\theta} r^2 \sin\theta \ \mathrm{d}r\mathrm{d}\theta\mathrm{d}\phi, \\
&= \frac{1}{2\pi}\sqrt{\frac{2l+1}{2}} \int_0^\infty \int_0^\pi \psi_{n,l}(r)\ P_l^0(\cos\theta)\ e^{ipr\cos\theta}\ r^2 \sin\theta \ \mathrm{d}r\mathrm{d}\theta.
\end{aligned} \tag{30}$$

Note that $\psi(p)$ needs to be normalized. Integrating over θ and ϕ variables, leads to,

$$\psi_{n,l}(p) = (-i)^l \int_0^\infty \frac{\psi_{n,l}(r)}{p}\ f(r,p)\,\mathrm{d}r. \tag{31}$$

Depending on l, this can be rewritten in following form (m' starts with 0),

$$\begin{aligned}
f(r,p) &= \sum_{k=2m'+1}^{m'<\frac{l}{2}} a_k \frac{\cos pr}{p^k r^{k-1}} + \sum_{j=2m'}^{m'=\frac{l}{2}} b_j \frac{\sin pr}{p^j r^{j-1}}, \quad \text{for even } l, \\
f(r,p) &= \sum_{k=2m'}^{m'=\frac{l-1}{2}} a_k \frac{\cos pr}{p^k r^{k-1}} + \sum_{j=2m'+1}^{m'=\frac{l-1}{2}} b_j \frac{\sin pr}{p^j r^{j-1}}, \quad \text{for odd } l.
\end{aligned} \tag{32}$$

The co-coefficients a_k, b_j of even- and odd-l states can be found in Tables 1 and 2 of [11].

3. Formulation of Information-theoretical Quantities

In this section we shall briefly discuss the various information-theoretic quantities along with their mathematical forms. This will provide the context where these quantities are defined and the relations between them.

3.1. Shannon Entropy (S)

Information is mobile and can be carried over from one place to another. The basic idea of information theory is that, the more one knows about an event, the less new information one is apt to get about it. If an event is very probable, there is much less uncertainty and thus it offers little new information. In summary, the information content is an increasing function of inverse of the probability of an event, and proportional to the uncertainty existing before its occurrence [73, 152, 153].

$$\text{Information received by occurrence of an event} = \text{Amount of uncertainty prevailed before its occurrence} \tag{33}$$

Shannon in 1948 connected the measure of information content with a discrete probability distribution $(p_1, p_2, p_3, ..., p_n)$ consisting of n different events, by giving a measure quantifying the uncertainty in results, in the given form [74],

$$S = -k \sum p_i \ln p_i, \tag{34}$$

where k is a positive constant depending on the choice of unit. This definition can be explained by choosing two limiting cases: (i) at first, when any of the $p_i = 1$ and others are *zero*, then for this certain event $S = 0$, which is minimum (ii) in case of equi-probability, where all $p_i = \frac{1}{n}$, and the uncertainty of the outcome is maximum, then S is also maximum ($S = k \ln n$) [152]. In essence, it can be said that, for a given distribution, lesser the probability of occurring an event, higher will be the uncertainty associated with it. Hence, after occurrence of that event, more information will come out. Potentially, S is claimed as the best measure of information [73].

The concept of S has been extensively used in wave mechanics to describe various phenomena, such as Colin conjecture [154, 155], atomic avoided crossing [156, 157], orbital-free DFT [158, 159], electron correlation [160, 161, 162,

163, 164], configuration-interaction, entanglement in artificial atoms [165], aromaticity [113] in many-electron systems, etc. Additionally, the Shannon entropy sum ($S_t = S_{\mathbf{r}} + S_{\mathbf{p}}$) which obeys well known BBM inequality and contains net information, can be proved to be a stronger version of the traditional Heisenberg uncertainty principle in quantum mechanics [73]. The entropic uncertainty relation has the mathematical form [73],

$$S_{\mathbf{r}} + S_{\mathbf{p}} \geq D(1 + \ln \pi), \tag{35}$$

where D refers to dimensionality of the system. Here, $S_{\mathbf{r}}$, $S_{\mathbf{p}}$ have following forms,

$$\begin{aligned}
S_{\mathbf{r}} &= -\int_{\mathcal{R}^3} \rho(\mathbf{r}) \ \ln[\rho(\mathbf{r})] \ d\mathbf{r} = 2\pi \left(S_r + S_{(\theta,\phi)}\right), \\
S_{\mathbf{p}} &= -\int_{\mathcal{R}^3} \Pi(\mathbf{p}) \ \ln[\Pi(\mathbf{p})] \ d\mathbf{p} = 2\pi \left(S_p + S_{(\theta,\phi)}\right), \\
S_r &= -\int_0^\infty \rho(r) \ \ln[\rho(r)] r^2 dr, \qquad \rho(r) = |\psi_{n,l}(r)|^2, \\
S_p &= -\int_0^\infty \Pi(p) \ln[\Pi(p)] \ p^2 dp, \qquad \Pi(p) = |\psi_{n,l}(p)|^2, \\
S_{(\theta,\phi)} &= -\int_0^\pi \chi(\theta) \ \ln[\chi(\theta)] \sin\theta d\theta, \qquad \chi(\theta) = |\Theta(\theta)|^2.
\end{aligned} \tag{36}$$

3.2. Rényi Entropy (R)

Rényi entropy is a one-parameter extension of family of information measures having several similar properties to S. It was introduced in 1961 as a generalized version of S which for a given probability distribution $P = (p_1, p_2, ..., p_n)$ [166] can be defined as,

$$R_\alpha = \frac{1}{1-\alpha} \ln \left(\sum_k p_k^\alpha \right). \tag{37}$$

It is to be noted that, R_α is considered as an alternative measure of information because (i) R_α is the exponential mean of information entropy whereas S provides the arithmetic mean of it, and (ii) at $\alpha \to \infty$, R_α reduces to S [73]. It is an information generating functional of α-order entropic moments, which can completely characterize density $\rho(\mathbf{r})$. In case of continuous density distribution, it may be expressible in terms of expectation values of density, in following standard form [166, 167],

$$R^{\alpha}[\rho(\mathbf{r})] = \frac{1}{(1-\alpha)} \ln \langle \rho(\mathbf{r})^{(\alpha-1)} \rangle, \quad 0 < \alpha < \infty, \ \alpha \neq 1. \tag{38}$$

Entropic uncertainty relations usually carry more information over the commonly used quantum mechanical uncertainty relations such as, $\Delta x \Delta p \geq \hbar$. While the latter signifies the area of a phase space ($\delta\mathbf{r}\delta\mathbf{p}$) obtained by resolution of measuring instruments, the former does not provide us with any such information. It suggests that with an increase of localization of the particle in phase space, the sum of uncertainties in position and momentum space escalates. The quantum mechanical uncertainty relation containing the phase space is of the form,

$$R_{\mathbf{r}}^{\alpha} + R_{\mathbf{p}}^{\alpha} \geq -\frac{1}{2}\left(\frac{\ln\,\alpha}{1-\alpha} + \frac{\ln\,\beta}{1-\beta}\right) - \ln\,\frac{\delta\mathbf{r}\delta\mathbf{p}}{\pi\hbar}; \quad \left(\frac{1}{\alpha}+\frac{1}{\beta}\right) = 2. \tag{39}$$

In the limit, when $\alpha \rightarrow 1$ and $\beta \rightarrow 1$, this uncertainty relation reduces to Eq. (35). However, the above relation presents a better insight about uncertainty as it contains all-order entropic moments [75]. But still further improvement is required, which remains an open challenging problem. Interestingly, Eq. (39) becomes sharper and sharper when the relative size of phase-space area $\left(\frac{\delta\mathbf{r}\delta\mathbf{p}}{\pi\hbar}\right)$ defined by the experimental resolution reduces. In fact, it *is the case*, as one enters more and more in to the quantum territory.

In quantum mechanics, R^{α} has been successfully employed to investigate and speculate many quantum phenomena such as, entanglement, communication protocol, correlation de-coherence, measurement, localization properties of Rydberg states, molecular reactivity, multi-fractal thermodynamics, production of multi-particle in high-energy collision, disordered systems, spin system, quantum-classical correspondence, localization in phase space [168, 169, 170, 171, 172, 173, 174, 175, 176, 177], etc.

Rényi entropies of order $\lambda(\neq 1)$ (λ is either α or β) are obtained by taking logarithm of λ-order entropic moment [167, 75]. In spherical polar coordinate these can be expressed in following simplified form, by means of some straightforward algebra [11],

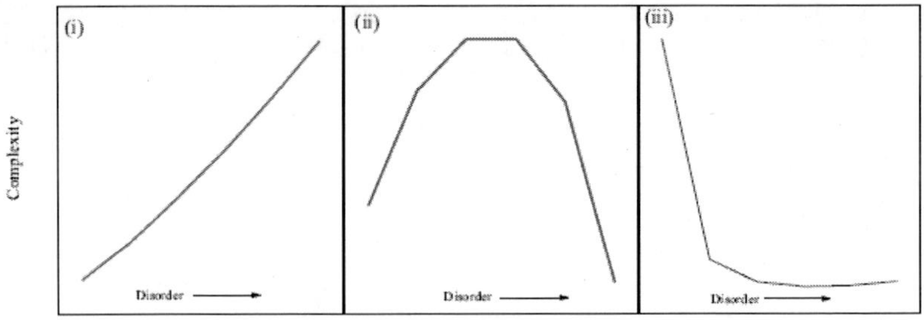

Figure 2. Three different categories of Complexities, as functions of disorder.

$$
\begin{aligned}
R_{\mathbf{r}}^{\lambda} = \frac{1}{1-\lambda}\ln\left(\int_{\mathcal{R}^3}\rho^{\lambda}(\mathbf{r})\mathrm{d}\mathbf{r}\right) &= \frac{1}{(1-\lambda)}\ln\left(2\pi\int_0^{\infty}[\rho(r)]^{\lambda}r^2\mathrm{d}r\int_0^{\pi}[\chi(\theta)]^{\lambda}\sin\theta\mathrm{d}\theta\right) \\
&= \frac{1}{(1-\lambda)}\left(\ln 2\pi + \ln[\omega_r^{\lambda}] + \ln[\omega_{(\theta,\phi)}^{\lambda}]\right), \\
R_{\mathbf{p}}^{\lambda} = \frac{1}{1-\lambda}\ln\left[\int_{\mathcal{R}^3}\Pi^{\lambda}(\mathbf{p})\mathrm{d}\mathbf{p}\right] &= \frac{1}{(1-\lambda)}\ln\left(2\pi\int_0^{\infty}[\Pi(p)]^{\lambda}p^2\mathrm{d}p\int_0^{\pi}[\chi(\theta)]^{\lambda}\sin\theta\mathrm{d}\theta\right) \\
&= \frac{1}{(1-\lambda)}\left(\ln 2\pi + \ln[\omega_p^{\lambda}] + \ln[\omega_{(\theta,\phi)}^{\lambda}]\right).
\end{aligned}
\tag{40}
$$

Here ω_τ^λs are entropic moments in τ (r or p or θ) space with order λ, having forms,

$$
\omega_r^{\lambda} = \int_0^{\infty}[\rho(r)]^{\lambda}r^2\mathrm{d}r, \quad \omega_p^{\lambda} = \int_0^{\infty}[\Pi(p)]^{\lambda}p^2\mathrm{d}p, \quad \omega_{(\theta,\phi)}^{\lambda} = \int_0^{\pi}[\chi(\theta)]^{\lambda}\sin\theta\mathrm{d}\theta.
\tag{41}
$$

3.3. Fisher Information (I)

The idea of entropy can satisfactorily explain the degree of disorder of a given phenomenon. Besides, however, it is required to find out a suitable measure of disorder whose variation derives the event. The concept of entropy is incapable of providing this. But, having the ability to approximate a parameter, I can serve as a good candidate for this purpose. Thus, it becomes a cornerstone of the statistical field of study, known as parameter estimation [178]. If e^2 be the mean-square error in an estimation of $\hat{\theta}$, then I measures the expected error in accordance with the following relation,

$$
e^2 I \geq 1; \quad I \geq \frac{1}{e^2}. \tag{42}
$$

Equation (42) suggests that, I is always greater than the reciprocal of e^2; only in case of *Gaussian* distribution it becomes equal to inverse of e^2. The general form of I is,

$$I = \int \frac{|\nabla \rho(\tau)|^2}{\rho(\tau)} \, \mathrm{d}\tau, \tag{43}$$

which is a gradient functional of density quantifying the local density fluctuation in a given space. While for a sharp distribution I is higher, it is lower for flat distribution. Thus, it is evident that with a rise in uncertainty, I decreases. It resembles the famous Weizsäcker kinetic energy functional, $T_\omega[\rho]$, frequently used in DFT [179, 158].

For central potential $I_{\mathbf{r}}$ and $I_{\mathbf{p}}$, the net Fisher information, in $\mathbf{r}$ and $\mathbf{p}$ spaces respectively, are expressed as [81],

$$\begin{aligned} I_{\mathbf{r}} &= \int_{\mathcal{R}^3} \left[\frac{|\nabla \rho(\mathbf{r})|^2}{\rho(\mathbf{r})}\right] \mathrm{d}\mathbf{r} = 4\langle p^2\rangle - 2(2l+1)|m|\langle r^{-2}\rangle \\ I_{\mathbf{p}} &= \int_{\mathcal{R}^3} \left[\frac{|\nabla \Pi(\mathbf{p})|^2}{\Pi(\mathbf{p})}\right] \mathrm{d}\mathbf{p} = 4\langle r^2\rangle - 2(2l+1)|m|\langle p^{-2}\rangle, \\ \langle r^2\rangle &= \int_0^{r_c} \psi^*_{n,l}(r) r^4 \, \psi_{n,l}(r)\mathrm{d}r, \quad \langle p^2\rangle = \int_0^{r_c} \psi^*_{n,l}(r)[-\nabla^2 \psi_{n,l}(r)] r^2 \mathrm{d}r \\ \left\langle \frac{1}{r^2}\right\rangle &= \int_0^{r_c} \psi^*_{n,l}(r)\psi_{n,l}(r)\mathrm{d}r, \quad \left\langle \frac{1}{p^2}\right\rangle = \int_0^{r_c} \psi^*_{n,l}(p)\psi_{n,l}(p)\mathrm{d}p. \end{aligned} \tag{44}$$

The above equations can be further modified in following equivalent forms [13, 9],

$$\begin{aligned} I_{\mathbf{r}} &= 8\mathcal{E}_{n,l} - 8\langle v(r)\rangle - 2(2l+1)|m|\langle r^{-2}\rangle \\ I_{\mathbf{p}} &= 8\mathcal{E}_{n,l} - 8\langle T\rangle - 2(2l+1)|m|\langle p^{-2}\rangle \\ &= \frac{8}{\omega^2}\langle v(r)\rangle - 2(2l+1)|m|\langle p^{-2}\rangle, \end{aligned} \tag{45}$$

where $v(p)$ is the p-space counterpart of $v(r)$.

When $m = 0$, $I_{\mathbf{r}}$ and $I_{\mathbf{p}}$ in Eq. (44) can be further simplified into,

$$I_{\mathbf{r}} = 4\langle p^2\rangle, \qquad I_{\mathbf{p}} = 4\langle r^2\rangle. \tag{46}$$

It is seen that, at fixed n, l, both $I_{\mathbf{r}}, I_{\mathbf{p}}$ provide maximum values when $m = 0$, and both of them decline with rise in m. Hence the upper bound for I_t can be obtained as,

$$I_{\mathbf{r}} I_{\mathbf{p}} \; (= I_t) \leq 16\langle r^2\rangle\langle p^2\rangle. \tag{47}$$

Further adjustment using Eq. (45) leads to following uncertainty relations [81],

$$\frac{81}{\langle r^2\rangle\langle p^2\rangle} \leq I_{\mathbf{r}} I_{\mathbf{p}} \leq 16\langle r^2\rangle\langle p^2\rangle. \tag{48}$$

Therefore, in a central potential, I-based uncertainty product is state dependent. It changes with variation in n, l and is bounded by both upper as well as lower limits.

3.4. Onicescu Energy (E)

In 1966, Onicescu introduced the concept of information energy (E), in an endeavour to provide a finer measure of dispersion distribution than that of S. E, for a discrete probability distribution can be defined as,

$$E = \sum_i p_i^2. \tag{49}$$

which can be extended for a continuous distribution as (expectation value of probability density),

$$E = \int \rho^2(\tau)\, \mathrm{d}\tau. \tag{50}$$

Therefore, in this occasion, disequilibrium and information energy have same form. This quantity shows a reverse trend to S; greater the information energy, more concentrated is the probability distribution and the information content reduces. Similar to the previous measures, it is also utilized in orbital-free DFT [159], testing normality [180], electron correlation [163], Colin conjecture [154, 155], configuration interaction [181] etc.

By definition, E refers to the 2nd-order entropic moment [179]; for central potential it assumes the form below (E_t is the Onicescu energy product),

$$\begin{aligned} E_r &= \int_0^\infty [\rho(r)]^2 r^2 \mathrm{d}r, \quad E_p = \int_0^\infty [\Pi(p)]^2 p^2 \mathrm{d}p, \\ E_{\theta,\phi} &= \int_0^\pi [\chi(\theta)]^2 \sin\theta \mathrm{d}\theta. \quad E_t = E_r E_p E_{\theta,\phi}^2. \end{aligned} \tag{51}$$

Uncertainty product for such measures are studied in [182].

3.5. Complexities

Simplified systems or idealizations are always a starting point to solve scientific problems. From the most basic grounds, based on our common knowledge, a system is said to be "complex" when it deviates from a pattern regarded as simple. An atom is itself a complex system and confinement introduces more complexity to it. Complexity arises in a system due to breakdown of certain symmetry rules. Quantitative study of this measure gives an idea of organization in a system and can be considered as a general descriptor of structure and correlation. Due to its dependence on the nature of description or on the scale of observation, no univocal characterization of this quantity is available. Some notable ones to be mentioned from the definitions available in literature are: Shiner, Davidson and Landsberg (SDL) [183, 184], Fisher-Shannon (C_{IS}) [119, 185, 186], Cramér-Rao [186, 100], generalized Rényi-like [187, 188, 98] complexity, etc. They can be divided into three broad categories, depending on their behavior: (i) advances monotonically with disorder (ii) reaches its minimal value for both completely ordered and disordered systems, and a maximum at some intermediate level (iii) increases with order. It has finite value in a state lying between two limiting cases of complete order (maximum distance from equilibrium) and maximum disorder (at equilibrium). Figure 2 pictorially depicts above three different kinds of complexities as functions of disorder.

The "statistical measure of complexity (C_{LMC})" is one such measure which is based on the statistical description of a system. In product form this can be written as, $C_{LMC} = H.D$ where H is nothing but the product of information content (such as S, R etc.) and D is the concentration of spatial distribution. This quantity was later criticized [189] and modified [190] to the form of $C_{LMC} = D.e^S$, to satisfy few conditions such as reaching minimal values for both extremely ordered and disordered limits, invariance under scaling, translation and replication. Interestingly, amongst the above mentioned complexities, C_{IS} corresponds to a measure which probes a system in terms of complementary global and local factors, and also satisfies certain desirable properties [179], like invariance under translations and re-scaling transformations, invariance under replication, near-continuity, etc. This has remarkable applications in the study of atomic shell structure, ionization processes [185, 186, 191], as well as in molecular properties like energy, ionization potential, hardness, dipole moment in the localization-delocalization plane showing chemically significant pattern [192], molecular reactivity studies [193]. Some elementary chemical reactions such as hydrogenic-abstraction reaction [194], identity SN^2 exchange reaction [195], and also concurrent phenomena occurring at the transition region [196] of these reactions have been investigated through composite information-theoretic measures in conjugate spaces.

Without any loss of generality, let us define complexity in following general form $C = Ae^{b.B}$. The order (A) and disorder parameters (B) may include (E, I) and (R, S) respectively. With this in mind, we are interested in the following four quantities,

$$C_{ER} = Ee^{bR}, \qquad C_{IR} = Ie^{bR}, \qquad C_{ES} = Ee^{bS}, \qquad C_{IS} = Ie^{bS}. \quad (52)$$

3.6. Relative Information

Kullback-Leibler divergence or relative entropy is a descriptor of deviation of a given probability distribution from a reference reference one [197, 198]. In quantum mechanics, this characterizes a measure of distinguishability between two states quantifying the change of information from one state to another [178]. Relative R and S were studied for various atoms using H atom ground-state as reference [199]. Their direct relation with atomic radii and quantum capacitance has been reported [99, 199]. Relative Fisher information (IR) is another such interesting measure [200]. It has witnessed considerable applications in physics and chemistry, such as, in the calculation of phase-space gradient of dissipated work and information [201], deriving Jensen divergence [202], relation with score function [203], in the context of probability current [204], in thermodynamics [205], etc. It has also been used in formulating atomic densities [206] and deriving density functionals, under local-density and generalized-gradient approximations [207]. Further, IR along with Hellmann-Feynman and virial theorem was used to develop a Legendre transform structure related to SE [208]. On the basis of estimation theory, it was designed in a self-consistent manner [209]. In quantum chemistry perspective, the above two theorems and entropy maximization principle were used in its formulation [210, 211, 212]. Very recently, IR for some exactly solvable potentials including 1D and 3D QHO in both position and momentum spaces was evaluated analytically using a ground state of definite symmetry (for example, $l = 0$ for s orbital, $l = 1$ for p orbital, and so on) as reference [12]. For H-atom using $1s$ orbital as basis, IR in position space was numerically estimated [213].

For two normalized probability densities $\rho_{n,l,m}(\tau)$, $\rho_{n_1,l_1,m_1}(\tau)$, IR can be written as,

$$\mathrm{IR}\left[\rho_{n,l,m}(\tau)|\rho_{n_1,l_1,m_1}(\tau)\right] = \int_{\mathcal{R}^3} \rho_{n,l,m}(\tau) \left|\nabla \ln\left\{\frac{\rho_{n,l,m}(\tau)}{\rho_{n_1,l_1,m_1}(\tau)}\right\}\right|^2 \mathrm{d}\tau. \quad (53)$$

Here n, l, m and n_1, l_1, m_1 are the identifiers of target and reference states respectively, while τ is a generalized variable. Without any loss of generality, in case of two central potentials, these probability densities $\rho_{n,l,m}(\tau)$, ρ_{n_1,l_1,m_1} can be conveniently expressed as,

$$\rho_{n,l,m}(\tau) = R^2_{n,l}(s)\Theta^2_{l,m}(\theta); \qquad \rho_{n_1,l_1,m_1}(\tau) = R^2_{n_1,l_1}(s)\Theta^2_{l_1,m_1}(\theta). \quad (54)$$

In the above equation, $R_{n,l}(s)$, $R_{n_1,l_1}(s)$ signify radial parts, where "s" implies either r or p variable and $\Theta_{n,l}(\theta)$, $\Theta_{n_1,l_1}(\theta)$ represent angular contributions of two wave functions. Thus Eq. (53) may be recast in the form,

$$\mathrm{IR}\left[\rho_{n,l,m}(\tau)|\rho_{n_1,l_1,m_1}(\tau)\right] = \mathrm{IR}\left[\rho_{n,l}(s)|\rho_{n_1,l_1}(s)\right] + \left\langle\frac{1}{s^2}\right\rangle \mathrm{IR}\left[\Theta^2_{l,m}(\theta)|\Theta^2_{l_1,m_1}(\theta)\right] +$$
$$2\int_0^\infty s\, R^2_{n,l}(s)\left[\frac{d}{ds}\ln\left\{\frac{R^2_{n,l}(s)}{R^2_{n_1,l_1}(s)}\right\}\right] \mathrm{d}s \int_0^\pi \Theta^2_{l,m}(\theta)\left[\frac{d}{d\theta}\ln\left\{\frac{\Theta^2_{l,m}(\theta)}{\Theta^2_{l_1,m_1}(\theta)}\right\}\right] \sin\theta \mathrm{d}\theta \quad (55)$$

where the following quantities have been defined,

$$\begin{aligned}
\mathrm{IR}\left[\rho_{n,l}(s)|\rho_{n_1,l_1}(s)\right] &= \int_0^\infty s^2\, \rho_{n,l}(s)\left|\frac{d}{ds}\ln\left\{\frac{\rho_{n,l}(s)}{\rho_{n_1,l_1}(s)}\right\}\right|^2 \mathrm{d}s \\
\mathrm{IR}\left[\Theta^2_{l,m}(\theta)|\Theta^2_{l_1,m_1}(\theta)\right] &= \int_0^\pi \Theta^2_{l,m}(\theta)\left|\frac{d}{d\theta}\ln\left\{\frac{\Theta^2_{l,m}(\theta)}{\Theta^2_{l_1,m_1}(\theta)}\right\}\right|^2 \sin\theta\, \mathrm{d}\theta
\end{aligned} \quad (56)$$

4. Result and Discussion

We will discuss the results in the order they were presented in Sec. 3. Thus first we analyze the information measures in DW potentials, followed by CHA and at last the atom-in-a-cage. In all cases, special attention has been paid to the excited states.

4.1. Double Well Potential

The particular form of DW potential, that has been chosen for the current exposition, has the form: $v(x) = \alpha x^m - \beta x^n + \gamma x^l$ (α, β, γ are positive real numbers, m, n, l are integers and $m > n > l$). Depending on the values of m, n, γ, it can fall into one of the two different classes (i) SDW, where m, n are *even* and $\gamma = 0$ (ii) ADW, where m, n *even*, $\gamma \neq 0$ and l *odd*. Thus it is apparent that, SDW may be regarded as a special case of ADW.

4.1.1. Symmetric Double Well Potential

The main concern is to understand the effect of β variation on the behavior of a particle in a SDW potential. An increment in β enhances both delocalization by increasing spacing between classical turning points, and also assists confinement through increase of barrier height. Therefore, an interplay between these two simultaneous and opposing effects should be felt in the pattern of information entropy as well–hence one or more extremum is to be expected. A detailed study of ground and first excited states reveals that, conventional uncertainty product $\Delta x \Delta p$ seems inadequate to explain such features. For both states Δx and $\Delta x \Delta p$ advance monotonically with rise in β. On the contrary, Δp first decreases, attains a minimum and then progresses with increment of β. But these trends cannot be used to interpret either trapping of the particle or tunneling [6].

Against this background, we begin our discussion by first following I_x, I_p, I. I_x for ground and first three excited states are qualitatively analogous; in all four cases it decreases with increase of β, attains a minimum and then progressively increases. For a particular state, these minima shift to right and gets more flattened with increase of α. Appearance of a minimum may be imposed to balance two competing effects. Two left panels (a), (d) of Fig. 4 (of ref. [6]) gather plots for I_x^0, I_x^1 and I_x^2, I_x^3 pairs with β, for a definite case of $\alpha = 1$. This clearly establishes the merging of two states after a particular β, which is undoubtedly due to a quasi-degeneracy in our SDW potential. This convergence point of I_x shifts to higher values of β with rise in α.

After observing the extremal nature in I_x, it is natural to explore the behavior of I_p or I. The qualitative nature of I_p and I appear to be quite similar. I_p progresses with advancement of β, but the rate of increase gets considerably slower for higher α. It holds true for all states considered. Also, one finds that, for all these states, at smaller β, progress in I_p seems rather nominal; it remains almost unaltered until a certain β is reached, after which I_p keeps on progressing drastically. The characteristic β at which this transition happens normally increases with state index. Moreover, for a certain state, this β is shifted to higher values as α advances. Importantly, however, unlike the case of I_x, no extremal nature is noticed in this context; instead one finds slight flatness for smaller β. This leads to the inference that, like traditional uncertainty measures, I_p also is incapable to sense the competitive effects in a SDW. However, as in I_x plots, here also, in two middle panels (b), (e) of Fig. 4 (of ref. [6]), (I_p^0, I_p^1) and (I_p^2, I_p^3) pairs convene at a particular β depending on a certain α value. This also could be a probable signature of quasi-degeneracy in this potential.

Next we present net fisher information I. In general, it accelerates with increase in β for all four states. Total information initially increases very slowly until a certain β is reached and after that it imprints drastic continuous growth. For a particular state, the β at which this transition occurs, is shifted right as α is increased. Progress of α consistently reduces growth rate of I in all four occasions. Like the I_p counterpart, I also fails in bearing any characteristic signature suggestive of the competing effects in these four states. However, as shown in two rightmost panels (c), (f) of Fig. 4 of ref. [6], pairs like (I^0, I^1) and (I^2, I^3) readily coalesce at a certain β. On the basis of above exploration, I is not decisive enough to explain the dual effects (delocalization and confinement), as well as tunneling.

Next we focus on S for lowest four states. Apparently, S_x at first increases steadily, then attains a maximum and finally falls off gradually with increasing β. Also, for a particular state, with increase of α the positions of these maxima shift to higher β and respective peak heights decrease. It infers that, at small β, delocalization effect prevails. However, at large β, the prevalent effect seems to be its confinement. The point of maximum signifies the characteristic β, at which maximum information can be extracted about that state of SDW. It is necessary to mention here that, change in α *does not* cause any significant alteration in the qualitative pattern of S_x–the maximum shifts to right, and S_x values get suppressed at maximum with serial increase in α. There appears a shoulder in S_x^2; its position also roughly coincides with onset of tunneling. Next, from panel (a) of Fig. 6 of ref. [6], it is noticed that S_x for ground and first excited state seem to coalesce at a value of β approximately close to 5; while same for second and third excited state in panel (d), occurs at closely $\beta = 9$. Both these are achieved at $\alpha = 1$. This merging of S_x is a signal of appearance of quasi-degeneracy in both pair of states, and also trapping of particle in either of the wells. Study of S_p and S will further consolidate this finding.

The general trend of S_p for ground and first three excited states seems to be quite similar, *viz.*, S_p initially decreases sharply with rise in β, then attains a minimum and continuously increases thereafter. In this case also, variation in

Table 1. Distribution of lowest six states within two wells, as well as number of effective nodes, for ADW potential, in Eq. (14). Four ranges of k are considered [7].

n	$0<k<1$		$1<k<2$		$2<k<3$		$3<k<4$	
	well	node	well	node	well	node	well	node
0	I	0	I	0	I	0	I	0
1	II	0	I	1	I	1	I	1
2	I	1	II	0	I	2	I	2
3	II	1	I	2	II	0	I	3
4	I	2	II	1	I	3	II	0
5	II	2	I	3	II	1	I	4

α shows no qualitative change. Positions of these minima switch to higher β with increasing α; however this time, values of S_p increase overall. A shoulder again appears for S_p^2. However, since effects of tunneling in momentum space is not clearly understood, a definitive explanation of this phenomenon appears difficult. In top middle segment (b) of Fig. 6 (of ref. [6]), S_p^0, S_p^1 corresponding to first two states of SDW potential are plotted against β, for a fixed $\alpha = 1$; analogous plots are drawn for S_p^2, S_p^3 in bottom middle panel (e). Note that, S_p's in case of a SDW show a style similar to that found in QHO. While these plots pertain to $\alpha = 1$, similar trend is recorded for other α as well, with a corresponding shift in location of convergence.

To study the cumulative effect of barrier on a given state, one needs to investigate net S. For ground state this quantity increases quite sharply, passes through an almost inconspicuous maximum and remains virtually constant thereafter. For first excited state, on the other hand, the same decreases monotonically, until individual α-plots merge together. Behavior of net S seems to be unaffected, at least qualitatively, with respect to variations in α–except for a shift in right. The maximum in it (second excited state) is much more distinctly defined than its counterpart in ground state. Apart from that, this maximum is now preceded by the appearance of a minimum. Now, for third excited state it initially follows same pathway as its counterpart in first excited state, falling steeply and monotonically. However, similarities end there; it then attains a minimum, rises rapidly and eventually decays into constancy.

Now, two rightmost panels (c), (f) of Fig. 6 of ref. [6] illustrate the convergence of net S for ground (S^0), first (S^1) and second (S^2), third (S^3) excited states successively, keeping α fixed at 1 in both instances. It becomes stationary at a value of 2.53 on and after $\beta = 5$ for first pair, while same for latter pair is attained at $\beta = 20$ with $S = 3.08$. However, these net values remain unaltered by variations in α. In all cases it obeys the bound given in Eq. (35).

Leaving aside the second excited state, E_x for other three states maintain qualitative agreement amongst themselves, i.e., E_x initially decreases, attains a minimum, then increases gradually. For a given state, positions of these minima are shifted to higher β with continuous increase of α. Also, the lowest value of E_x increases and these minima get flattened with progress in α. E_x^2 is again different from remaining three states, where one observes a shoulder before reaching minimum and then increases monotonically. The position of this shoulder and minima get shifted to right with increase in α. It clearly demonstrates the marked contrast between second excited state from remaining three states; former shows a sequence of extrema while latter three are characterized by a minimum. This uniqueness in $n = 2$ could be imposed to the simultaneous effects of β on particle–at small values, it indicates the increase of delocalization of particle; however at large values, the deciding effect seems to be its confinement. Such kink has been observed earlier in case of S_x^2. Note that, position of shoulder also roughly coincides with onset of tunneling. Further, from two left segments (a), (d) of Fig. 10 of ref. [6], it is transparent that, E_x for lowest two states merge at a value of $\beta \approx 5$ for $\alpha = 1$, whereas for second, third excited states, convergence occurs at $\beta \approx 9.5$ for same α. This joining of E_x's is a clear suggestion of occurrence of quasi-degeneracy in both pairs of states, and also confinement of particle in either of the wells.

Akin to E_x, in p space, the general trend for ground, first and third excited state maintains a qualitative similarity amongst themselves; i.e., E_p initially advances sharply with increase in β, reaches a maximum and gradually decreases afterwords. Once again, variation in α shows no qualitative change here. Positions of maxima shift to higher β and get flatter with rise in α. variation of E_p^2 with β is, however, substantially different and more interesting; one observes two maxima sandwiched by a minimum in E_p^2. At first, E_p^2 increases attaining a maximum, then drops down to a minimum, again increases to a maximum and finally decreases. Clearly, for progress in α, they get right-shifted and individual plots flatten. Again, this effect may be due to quasi-degeneracy in this potential. Next, two middle panels (b), (e) of Fig. 10 of ref. [6] depict E_ps for two lowest pair of states, against β. They meet at nearly $\beta = 7.5$ and 11.75. Note that while these plots pertain to $\alpha = 1$, a similar motif is noticed for other α as well, with a resembling shift in location of convergence point.

In case of ground state net E quite sharply decreases initially to attain a minimum, then almost immediately hits a maximum, again reduces to reach a constant (0.1195), for all α. Locations of both minima and maxima shift towards right for higher α. Moreover, distance between these two extrema enhances with increase of α. In contrast, for first excited state it increases monotonically and converge together at 0.1195. This situation modifies for second, third excited states; in former case, it initially rises to reach a maximum, then follows through a minimum and again increases until attains a constant (0.0667). The extrema, in this case, are much more distinct and well separated than its counterpart in ground state. Values of net E at extrema are quite comparable for all α studied. Increase of α transfers the extrema towards right but all of them eventually converge to constant value of 0.0667. For third excited state, it increases to reach a maximum, then gradually decreases to attain a constant (0.0667). Now, top right panel (c) of Fig. 10 of [6] visualizes convergence of ground and first excited state; bottom right panel (f) does same for 2nd and 3rd excited state, keeping α constant at 1 in both occasions. Net E becomes stationary at a value of 0.1195 on and after $\beta = 5$ for first pair, while same for latter pair happens at $\beta = 20$ with net $E = 0.0667$.

Next, we study the Onicescu-Shannon complexity measure (symbolized as OS) for first four states of SDW potential. Except for second excited state, OS_x decreases with β, proceeds through a minimum and sharply rises thereafter. Positions of these minima for all these three states shift towards right as α is increased. This is again possibly due to competing effects in a SDW. OS_x^2 behaves quite differently from remaining three states. Thus, initially there appears a shallow minimum followed by a small maximum and finally a global minimum. With α, however, positions of these extrema switch towards right as in other three states. Further, segments (a), (d) of Fig. 13 (of ref. [6]) establish that the pairs OS_x^0, OS_x^1 and OS_x^2, OS_x^3 convene at a certain β, indicating appearance of quasi-degeneracy. In this instance also, qualitative behavior of OS_p for ground, first and third excited state are analogous, while second excited state stands out. OS_p for former three states are characterized by a prominent maximum followed by a gradual decrease, while the lone second state exhibits a maximum and minimum in succession. Positions of the maxima switches to higher β as α increases. As usual, the pairs OS_p^0, OS_p^1 and OS_p^2, OS_p^3 smoothly converge at a particular β in middle column in (b), (e) of Fig. 13 (of ref. [6]).

OS^0 at first, falls down to a minimum slowly, then grows to a maximum

Table 2. $I_{\mathbf{r}}$, $I_{\mathbf{p}}$ for first five s orbitals of CHA, at seven different r_c [13]. See text for details

n	$r_c = 0.1$	$r_c = 0.5$	$r_c = 1$	$r_c = 5$	$r_c = 10$	$r_c = 20$	$r_c = 100$
				$I_{\mathbf{r}}^{\dagger}$			
1	3948.73709	158.896112	40.5850917	4.1962752	4.00009944	4.00000000	4.00000000
2	15791.8212	632.093249	158.322897	6.14412880	1.48824976	1.00148803	1.00000000
3	35530.8589	1421.49254	355.555298	14.2086744	3.31621534	0.70203512	0.44444444
4	63165.6652	2526.80603	631.828614	25.2980567	6.18599632	1.37620355	0.25000000
5	98696.1908	3947.98150	987.090739	39.5169227	9.79535706	2.31004148	0.1600061
				$I_{\mathbf{p}}^{\ddagger}$			
1	0.01119745	0.26851341	1.01251135	10.7398867	11.9978379	11.9999999	12.0000000
2	0.01284003	0.32261837	1.30128804	35.6220106	114.097280	167.397282	168.000000
3	0.01312184	0.32950852	1.32621209	35.4490641	152.667868	568.122689	827.999999
4	0.01321715	0.33151636	1.33186010	34.7398719	146.840310	640.181092	2591.99999
5	0.01326029	0.33232931	1.33360252	34.3394100	142.869003	610.819403	6299.58443

$^{\dagger}I_{\mathbf{r}}$ of FHA for $n = 1 - 5$ are: 4, 1, $\frac{4}{9}$, 0.25, 0.16.
$^{\ddagger}I_{\mathbf{p}}$ of FHA for $n = 1 - 5$ are: 12, 168, 828, 2592, 6300.

very sharply, eventually becoming constant. These extrema shift towards higher β with increasing α; however, they finally converge to same constant (0.6465). OS^1 initially increases with β, then converges to a stationary value of 0.6465. Rate of increase of OS^1 with β declines with increase of α. OS^2 initially rises to a maximum, then follows through a minimum and again convenes to a constant (0.525). All these extrema move toward right with increase of α. OS^3 rises to a maximum and converges to same value (0.525). In this case also, a progress in α leads to same result of shifting maxima towards right. Additionally, in two rightmost panels (c), (f) of Fig. 13 of ref. [6], we provide convergence of OS for first and second pairs respectively. More elaborate works of information measure of SDW is available in [6].

4.1.2. Asymmetric Double Well Potential

An ADW potential is invoked by adding a linear term in SDW potential. This asymmetric term (γx) immediately lifts the quasi-degeneracy in energy. However, a new kind of degeneracy emerges at certain characteristic γ. Extensive study of energy exposes that, at each α there appears a range of γ after which energy increases steadily. It has been found that, for a fixed α, there exists a *positive* real number $k = \frac{\gamma}{\Delta\gamma}$, depending upon which, following rules for quasi-degeneracy in energy has been proposed [7].

(i) When $(n \geq k)$: three possible outcomes can be envisaged in this scenario.

(a) *k is odd positive integer*: an *odd-n* state will be quasi-degenerate with its immediate higher state.

(b) *k is even positive integer*: an *even-n* state will be quasi-degenerate with its adjacent higher state.

(c) *k is a fraction*: no quasi-degeneracy is possible.

(ii) When $k > n$: n-th state *can not* be quasi-degenerate. However, other higher ($> n$) states may show degeneracy depending on k, as delineated above in (i.a), (i.b).

These quasi-degeneracy rules were demonstrated in [7]. A in-depth analysis also tells that, at a definite α, by controlling β, γ desired quasi-degenerate pair can be generated. Examination of wave functions and densities for various states [7] uncovers that, a particle in ground state always resides inside deeper well. On the other hand, particle in an arbitrary excited state oscillates between deeper and shallower wells, and eventually gets trapped inside deeper well. This observation has been quantified in terms of some simple rules to forecast localization/delocalization of a particle in a certain well, in terms of the parameter $k(= \frac{\gamma}{\Delta\gamma})$ and state index n. For $\mathcal{E}_n < V_0$, following situations could be visualized.

(i) $n \geq k$: two possibilities arise:

(a) k is integer: particle in nth state is distributed in both well I and well II.

(b) k is fraction: four possibilities need to be considered:

(1) both n and integer part of k *even*: particle stays in well I.

(2) both n and integer part of k *odd*: particle stays in well I.

(3) n *even* and integer part of k *odd*: particle stays in well II.

(4) n *odd* and integer part of k *even*: particle stays in well II.

(ii) $n<k$: particle always resides in larger (I) well.

It is worthwhile making some remarks about nodes in wave functions. Table 1 provides number of effective nodes present in a particular state, at same selected range of k. In fact, this increases with state index (nth state has n nodes lying within classical turning points). However, because of occurrence of two separate wells as discussed above, certain nodes become insignificant. In range $0 < k < 1$, for example, $n = 0{,}1$ are practically node-less as they behave as ground state of wells I, II respectively. Analogously, $n = 2{,}3$ possess single node, for they represent first excited state of wells I, II; whereas $n = 4{,}5$ contain two nodes corresponding to second excited states of wells I, II respectively. These numbers are provided in third column of this table. Next, fifth column concludes that, within $1 < k < 2$, $n = 0{,}2$ have zero node; thus they represent the lowest two states of wells I, II. Also, $n = 1{,}4$ have single node, as they appear as first excited states of wells I, II. And $n = 3{,}5$ possess 2,3 nodes respectively, for they relate to second and third excited states of well I. In a similar fashion, in range $2 < k < 3$, $n = 0{,}1{,}2{,}4$ behave as lowest four states of well I, whereas $n = 3{,}5$ act as lowest two states of well II. These are clear from column seven. Finally for range $3 < k < 4$, last column ensures that $n = 0{,}1{,}2{,}3{,}5$ serve as first five states of well I, giving rise to 0,1,2,3,4 nodes respectively, whereas $n = 4$ remains effectively node-less, as it becomes the ground state of well II. Thus, one concludes that at fractional k, wells I and II virtually function as two different potentials (after a certain threshold β).

We now move on to information measures for first four energy states. Figure 9 of ref. [7] imprints S_x as a function of γ at three β (5, 10, 15) in top three rows, A(a)-A(c). Increase in β only makes the particle more trapped within well I or II. Generally, at certain γ, plot for nth state quickly jumps to a crest and then gradually decreases by traveling through $n + 1$ peaks. After that, particle ultimately localizes in larger (I) well. These jumps at even γ (integer k) convey transition points, which occurs only at fixed interval of $\Delta\gamma$, and observed up to a certain γ (characteristics for a given state). At fractional k, development in S_x is rather small because of confinement. Once, it finally rests in larger well ($k > n$), S_x decreases with rise in γ explaining gradual build-up of confinement. Next B(a)-B(c), delineate that slope of S_p against γ curves varies noticeably at intervals of $\Delta\gamma$. This discontinuity indicates the transition of particle from one well to another. Again, an nth state leads to (n+1) jumps in S_p before localization occurs in well I. Progress of net S as a function of γ are shown in bottom row. Like S_x, net S also portrays pronounced peaks at intervals of $\Delta\gamma$ indicating transition points at integer k. Apart from that, like S_x, an nth state gives rise to (n+1) peaks in S_p before settling in well I. When particle resides in a specific well, extent of confinement is not always same. Visibly, when $0<k<1$ $(0<\gamma<2)$, net S for $n=0,1$ remain very intimate. This is narrated from the fact that, at this interval these two states act effectively as lowest state of wells I, II. Likewise, for $1<k<2$ $(2<\gamma<4)$, net S of $n=0,2$ are closer to each other, as they behave as lowest states of wells I, II. As anticipated, in $2<k<3$, net S of $n=0,3$ approach each other quite closely. This observations could be explained from a consideration of number of effective nodes in Table 1 in each of these states in respective interval of γ. Thus, at nth interval of $\Delta\gamma$, nth state behaves similar to ground state of a definite well. Note that there is a jump in S_x^0, S_p^0, S^0 at $\gamma=0$, informing a critical transition point, after which a particle remains in well I. This also mirrors the reality that, SDW is a special case of ADW potential.

We next examine progress of S with β through four lowest states taking γ as parameter. In all cases S_x initially rises with β before attaining a maximum and then declines asymptotically. This arises due to the dual effect of β. Interestingly, positions of these maxima move to left with increase in γ, for any given state; moreover they tend to vanish with γ, with higher state requiring larger γ. This is illustrated by considering the predominance of asymmetry over competing effect due to localization of particle in a definite well. Depending upon odd, even values of γ, two different tendency in S_x is seen beyond a certain threshold β. For $k=0$ $(\gamma=0)$, S_x of $n=0$,1 are very close; similarly at $k=1$ $(\gamma=2)$, same happens for $n=1$,2; at $k=2$ $(\gamma=4)$, $n=2$,3 are close. Continuing in this fashion, one can predict that, S_x of $n=3$,4 will approach each other at $k=3$ or $\gamma=6$ (not shown) and so on. This closeness of various sets of S_x at certain k is in consonance with quasi-degeneracy rule. On the other hand, when γ is odd (1, 3, 5, $\cdots$), then at $k=0.5, 1.5, 2.5$, S_x of $n=0$ becomes closer to $n=1$,2,3 states respectively. This, again is due to number of effective node of those states becoming equal at those specific k.

One notices that, generally, S_p declines with rise in β, attains a minimum and then increases gradually. Again, positions of these minima move towards left with progress in β. Beyond a particular γ, the extrema tend to die out. Also for higher n, there exist some occasional humps in S_p. Alike to S_x, S_p in certain states also converge depending on odd, even γ. However, the alterations of net S with β is not straightforward; complicated sequence of maxima, minima is

Table 3. $R_{\mathbf{r}}^{\alpha}$, $R_{\mathbf{p}}^{\beta}$ and $R^{(\alpha,\beta)} = (R_{\mathbf{r}}^{\alpha} + R_{\mathbf{p}}^{\beta})$ for lowest two s states in a CHA, at various r_c, for $\alpha = \frac{3}{5}, \beta = 3$ respectively. See text for more details

1s				2s			
r_c	$R_{\mathbf{r}}^{\alpha}$	$R_{\mathbf{p}}^{\beta}$	$R^{(\alpha,\beta)}$	r_c	$R_{\mathbf{r}}^{\alpha}$	$R_{\mathbf{p}}^{\beta}$	$R^{(\alpha,\beta)}$
0.1	−6.04495302	12.254494	6.209541	0.1	−6.06527856	14.246181	8.180902
0.2	−3.97406865	10.182673	6.208604	0.2	−3.98570108	12.160542	8.174841
0.5	−1.25276392	7.458575	6.205811	0.5	−1.23590502	9.387514	8.151609
1.0	0.77359587	5.427664	6.201260	1.0	0.84696859	7.245474	8.092443
5.0	4.66206639	1.524658	6.186724	5.0	5.76615415	1.215531	6.981685
7.5	4.93915225	1.263505	6.202657	7.5	6.96559616	−0.391676	6.573919
10.0	4.97268114	1.238720	6.211402	10.0	7.70531072	−1.298410	6.406900
40.0	4.97592206	1.237321	6.213243	40.0	8.61299696	−2.335383	6.277613

observed for different states for various β.

Similar study was performed for development of I, E, as well as Onicescu-Shannon information measures with changes in γ and β. A careful analysis unveils that, nature of I_x, E_x, OS_x with increase of γ are qualitatively similar to those of S_p. Likewise, trends of I_p, E_p, OS_p with rise in γ resemble the behavior of S_x. The conclusions from β variation are also in harmony to S. Overall all these measures lead to comparable conclusions as obtained through conventional uncertainty product, and S in position and momentum space.

4.2. Hydrogen Atom

4.2.1. Information Analysis for CHA

At the beginning it is appropriate to point out that, in case of central potential *net* information measures in both r and p spaces are separated into radial and angular parts. In a given space, the results provided resemble that of net measure including angular part. An FHA can be transformed to CHA by shifting the boundary from *infinity* to finite region. This variation in radial environment does not influence the angular boundary conditions. Hence, the angular contribution of information measure in FHA and CHA remain unaltered in both r and p spaces. For calculation of S, R, E magnetic quantum number m has been kept fixed to 0. However, I has been investigated for non-zero m states. The radial wave functions in r and p spaces depend only on n, l quantum number.

Hence, in both spaces radial wave function can be achieved by taking $m = 0$. Apart from that, a change in m from *zero* to *non-zero* value does not affect the expression of radial wave function in p space.

At first, $I_{\mathbf{r}}, I_{\mathbf{p}}$ are computed from Eq. (44). In this equation, angular part is normalized to unity. Hence, estimation of these derived quantities by employing only radial part serves the purpose. In order to understand the effect of confinement using I, pilot calculations are done by the authors in [13], for $1s, 2s, 2p, 3s, 3d, 4s, 4f$ and all l-states corresponding to $n = 5$ and 10, varying r_c from 0.1 to 300 a.u. This chapter discusses some of these results.

As discussed before, I is a measure of fluctuation in a given probability distribution. The analytical forms of $I_{\mathbf{r}}$ and $I_{\mathbf{p}}$ in a FHA were given in [81],

$$I_{\mathbf{r}}(Z) = \frac{4Z^2}{n^2}\left[1 - \frac{|m|}{n}\right], \tag{57}$$

$$I_{\mathbf{p}}(Z) = \frac{2n^2}{Z^2}\left[(5n^2 + 1 - 3l(l+1)) - |m|(8n - 6l - 3)\right]. \tag{58}$$

Equation (57) endorses that, at a certain m, $I_{\mathbf{r}}(Z)$ declines with progress in n and for a given n, linearly decreases with growth in m. However, it remains unaffected with change in l. At $m = 0$, $I_{\mathbf{r}} = 4\langle p^2 \rangle = 8 \times KE$ (KE is the average kinetic energy). On the other hand, at a definite l, m, $I_{\mathbf{p}}(Z)$ rises with n; but decreases with advancement of l for fixed n, m values. Similarly, for a certain n, l, $I_{\mathbf{p}}(Z)$ regresses with enhancement of $|m|$.

A careful analysis suggests that the trends of $I_{\mathbf{r}}, I_{\mathbf{p}}$ in CHA are not always in consonance with FHA. Table 2 imprints the behavior of $I_{\mathbf{r}}, I_{\mathbf{p}}$ for first five s orbitals of CHA at seven different r_c $(0.1, 0.5, 1, 5, 10, 20, 100)$. $I_{\mathbf{r}}$ for a definite n, l, m state declines with enhancement in r_c. In FHA, $KE = \frac{Z^2}{n^2}$ decreases with rise in n. On the contrary, in case of CHA, at a fixed r_c, KE increases with progress in n. This trend modifies at higher r_c. $I_{\mathbf{p}}$ at a fixed n, l, m state enhances with growth in r_c. However, unlike $I_{\mathbf{r}}$, pattern of $I_{\mathbf{p}}$ with change of n is straightforward. At low r_c region, it decreases with n (at fixed l, m). For all these states, $I_{\mathbf{r}}, I_{\mathbf{p}}$ eventually converge to respective FHA limits.

At fixed n, l, r_c, the effect of m quantum number on $I_{\mathbf{r}}, I_{\mathbf{p}}$ is candid. Like FHA, both $I_{\mathbf{r}}, I_{\mathbf{p}}$ in CHA decrease with increment in $|m|$. A sample result for $10k$ state is given in panels (a) and (b) of Fig. 2 (of ref. [13]) for $I_{\mathbf{r}}, I_{\mathbf{p}}$ respectively, as a function of r_c at five different $|m|$, namely $0, 2, 4, 6, 7$. It is interesting to mention that, $I_{\mathbf{p}}$ for $l \neq 0$ states having $m = 0$, increases with progress in r_c and finally reaches to their FHA values. However, for *non-zero*

Table 4. $S_{\mathbf{r}}$, $S_{\mathbf{p}}$, S for $1s$, $2s$ states in CHA at some chosen r_c. S's for all these states obey the lower bound given in Eq. (35). See text for details

1s				2s			
r_c	$S_{\mathbf{r}}$	$S_{\mathbf{p}}$	$S = S_{\mathbf{r}} + S_{\mathbf{p}}$	r_c	$S_{\mathbf{r}}$	$S_{\mathbf{p}}$	$S = S_{\mathbf{r}} + S_{\mathbf{p}}$
0.1	−6.24450338	12.8535	6.6089	0.1	−6.44745791	14.638	8.1905
0.2	−4.17785640	10.7787	6.6008	0.2	−4.36923353	12.5593	8.1900
0.5	−1.47034068	8.0472	6.5768	0.5	−1.62307869	9.8112	8.1881
1.0	0.52903030	6.0114	6.5404	1.0	0.45546229	7.7347	8.1901
5.0	4.01744418	2.524361	6.541805	5.0	5.46416082	2.8173	8.2814
7.5	4.13932453	2.425508	6.564832	7.5	6.72302624	1.3022	8.025
10.0	4.14460143	2.421936	6.566538	10.0	7.44615626	0.2765	7.7226
40.0	4.14472988	2.421862	6.566592	40.0	8.11092936	−0.75758	7.35334

l, m (at a fixed n) states, it passes through a maximum before reaching FHA limit.

It is well-known that, $I_{\mathbf{r}}$ in FHA is independent of l quantum number. However, in CHA, at a certain n, m, r_c it decreases with increment of l. One such example is panel (a) of Fig. 3 (of ref. [13]), where, $I_{\mathbf{r}}$ is plotted against r_c for $5s - 5g$. This graph complements above conclusion about $I_{\mathbf{r}}$; at low r_c region it reduces with l and at $r_c \rightarrow \infty$, becomes invariant of l. At a certain $n, |m|, r_c$, $I_{\mathbf{p}}$ in CHA advances with rise in l. However, in FHA it exerts a complete opposite trend, reducing with growth in l. Hence, there exists a crossover region in r_c after which the behavioral pattern of $I_{\mathbf{p}}$ alters to match the respective FHA. Panel (b) of Fig. 3 (of ref. [13]) establishes this, where $I_{\mathbf{p}}$ for $5s - 5g$ are pictorially given as functions of r_c.

The authors have investigated S, R, E in both r, p spaces for various states of CHA [11, 94]. Here we take this opportunity to address some of the interesting outcomes of the above exploration. Let us begin with R. In case of all n, l, m states of CHA, $R_{\mathbf{r}}^{\alpha}$ increases with rise in r_c. Conversely, $R_{\mathbf{p}}^{\beta}$ for all these states declines with growth in r_c. Table 3 imprints the values of $R_{\mathbf{r}}^{\alpha}$, $R_{\mathbf{p}}^{\beta}$ and $R^{(\alpha,\beta)}$ for $1s$ and $2s$ states at eight selected r_c values. Apart from that, it is important to mention that, at *small* r_c, Rényi entropies for $2p, 3d, 4f, 5g$ states in r, p spaces obey same order: $R_{\mathbf{r}}^{\alpha}(5g) > R_{\mathbf{r}}^{\alpha}(4f) > R_{\mathbf{r}}^{\alpha}(3d) > R_{\mathbf{r}}^{\alpha}(2p)$ and $R_{\mathbf{p}}^{\beta}(5g) > R_{\mathbf{p}}^{\beta}(4f) > R_{\mathbf{p}}^{\beta}(3d) > R_{\mathbf{p}}^{\beta}(2p)$. But at $r_c \rightarrow \infty$ limit, ordering in p space completely reverses, while the r-space ordering is preserved.

Dependence of R on n is quite interesting. At a constant l and $r_c = 0.1$,

$R_{\mathbf{r}}^{\alpha}$ declines with rise in n, while $R_{\mathbf{p}}^{\beta}$ and $R_{t}^{\alpha,\beta}$ show reverse tendency. This has been pictorially represented in bottom panels A(a), B(a), C(a) in Fig. 8 (of ref. [9]) respectively. This clearly explains that, effect of confinement enhances with progress in n. In this regard, it is worthwhile mentioning that, at $r_c \to \infty$, exactly opposite pattern with respect to $R_{\mathbf{r}}^{\alpha}$ and $R_{\mathbf{p}}^{\beta}$ is seen. First two columns of middle and top rows of Fig. 8 (of ref. [9]) indicate the appearance of maxima and minima in the plots in r, p spaces respectively. This suggests that with increase in r_c, the system approaches towards FHA. These extrema will shift to right with rise in r_c. Now, regarding the dependence of R on l, at a fixed n, $R_{\mathbf{r}}^{\alpha}$ initially decreases to attain a minimum and then increases with rise in l. However, $R_{\mathbf{p}}^{\beta}$ under same condition (fixed n), decreases with progress in l. This has been thoroughly discussed in [94] for $n = 10$ states.

Now we move on to S. Akin to $R_{\mathbf{r}}^{\alpha}$ and $R_{\mathbf{p}}^{\beta}$, $S_{\mathbf{r}}$ and $S_{\mathbf{p}}$ for a given n, l, m state progresses and declines with rise in r_c. Table 4 displays $S_{\mathbf{r}}$, $S_{\mathbf{p}}$, S_t for $1s$ and $2s$ states at eight selected r_c. Moreover, $S_{\mathbf{r}}$ imprints separate pattern to $R_{\mathbf{r}}^{\alpha}$ but $S_{\mathbf{p}}$ depicts similar behavior favouring to $R_{\mathbf{p}}^{\beta}$ in smaller r_c. The observed nature for $2p, 3d, 4f, 5g$ in r, p spaces is slightly unexpected: $S_{\mathbf{r}}(4f) > S_{\mathbf{r}}(5g) > S_{\mathbf{r}}(3d) > S_{\mathbf{r}}(2p)$ and $S_{\mathbf{p}}(5g) > S_{\mathbf{p}}(4f) > S_{\mathbf{p}}(3d) > S_{\mathbf{p}}(2p)$ [94]. As seen before, at $r_c \to \infty$ the order in r, p spaces reorganize to $S_{\mathbf{r}}(5g) > S_{\mathbf{r}}(4f) > S_{\mathbf{r}}(3d) > S_{\mathbf{r}}(2p)$ and $S_{\mathbf{p}}(5g) < S_{\mathbf{p}}(4f) < S_{\mathbf{p}}(3d) < S_{\mathbf{p}}(2p)$.

Our next objective is to understand the dependence of S on n. At a constant l and $r_c = 0.1$, $S_{\mathbf{r}}$, $S_{\mathbf{p}}$ and S_t resemble that of $R_{\mathbf{r}}^{\alpha}$, $R_{\mathbf{p}}^{\beta}$ and $R_{t}^{\alpha,\beta}$ respectively. The former decreases and latter two increase with rise in n. This has been exhibited in bottom panels A(a), B(a), C(a) of Fig. S2 (supporting document of ref. [9]) successively. This again explains the fact that, effect of confinement accelerates with advancement in n. It is needless to mention that, at $r_c \to \infty$, exactly reverse manner for $S_{\mathbf{r}}$ and $S_{\mathbf{p}}$ is observed. Here again, first two columns of middle and top rows of Fig. S2 (available in the supporting document of ref. [9]) suggest the appearance of maxima and minima in the plots in r, p spaces respectively. This suggests that, with relaxation in confinement the system approaches towards FHA. These extrema will move towards right with rise in r_c. We can now follow the effect of l on S. Similar to $R_{\mathbf{r}}^{\alpha}$, here also at a constant n, $S_{\mathbf{r}}$ initially falls down to a minimum and then increases with progress in l. Parallel to $R_{\mathbf{p}}^{\beta}$, $S_{\mathbf{p}}$ at a fixed n, declines with growth in l. This has been discussed in detail [94] using $n = 10$ states as reference.

E exerts opposite effect to that of R, S. In case of a particular n, l, m state, $E_{\mathbf{r}}$ diminishes and $E_{\mathbf{p}}$ escalates with enhancement in r_c. Table 5 presents

Table 5. $E_{\mathbf{r}}, E_{\mathbf{p}}$ for $1s$ and $2s$ states in CHA at some chosen r_c. See text for details

$1s$			$2s$		
r_c	$E_{\mathbf{r}}$	$E_{\mathbf{p}}$	r_c	$E_{\mathbf{r}}$	$E_{\mathbf{p}}$
0.1	685.24426269	0.00000395	0.1	1467.68253819	0.0000005
0.2	87.40227398	0.00003142	0.2	185.27985826	0.0000045
0.5	5.97242136	0.00047889	0.5	12.20852686	0.0000715
1.0	0.84791755	0.00364537	1.0	1.59792065	0.0005811
5.0	0.04217592	0.16706123	5.0	0.01264653	0.1756848
7.5	0.03985512	0.20591605	7.5	0.00303307	1.0029980
10.0	0.03978990	0.20886414	10.0	0.00135661	2.6889282
40.0	0.03978873	0.20897494	40.0	0.00077712	7.6497493

$E_{\mathbf{r}}, E_{\mathbf{p}}, E_t$ for $1s, 2s$ at eight r_c, of CHA. As r_c leads to zero, $E_{\mathbf{r}}$ follows the ordering $E_{\mathbf{r}}(5g) > E_{\mathbf{r}}(4f) > E_{\mathbf{r}}(3d) > E_{\mathbf{r}}(2p)$ which converses to $E_{\mathbf{r}}(2p) > E_{\mathbf{r}}(3d) > E_{\mathbf{r}}(4f) > E_{\mathbf{r}}(5g)$ in $r_c \rightarrow \infty$ limit. Contrariwise, $E_{\mathbf{p}}$ imprints exactly opposite feature from its r-space counterpart at both small and large r_c regions. Now we can discern the reliance of E on n. At a certain l and $r_c = 0.1$, $E_{\mathbf{r}}$, $E_{\mathbf{p}}$, E_t assert completely opposite tendency to what is seen in R, S. The former advances and latter two decline with rise in n. This is demonstrated in bottom panels A(a), B(a), C(a) of Fig. S3 (of supporting document of ref. [9]) correspondingly. This again appreciates the fact that, effect of confinement prevails with advancement in n. It is appropriate to mention that, at $r_c \rightarrow \infty$, exactly inverse nature for $E_{\mathbf{r}}$ and $E_{\mathbf{p}}$ is attained. Here again, first two columns of middle and top rows of Fig. S3 of supporting document of ref. [9] suggest the appearance of maxima and minima in the graphs in r, p spaces respectively. This supports that, with relaxation in confinement, CHA proceeds towards FHA. These extrema will move forward towards right with rise in r_c. The effect of l on E is completely different to $R_{\mathbf{r}}^{\alpha}, S_{\mathbf{r}}$. At a constant n, $E_{\mathbf{r}}$ abates and $E_{\mathbf{p}}$ escalates with progress in l. This has been discussed at length [94] in the context of $n = 10$ states.

Four different complexity measures, given in Eq. (52) are probed with change of r_c, choosing two different and commonly used values of b $(1, \frac{2}{3})$. Note that, for $b = 1$, $C_{ES}^{(2)}$ modifies to C_{LMC}; analogously $C_{IS}^{(1)}$ coincides with C_{IS} at $b = \frac{2}{3}$. In order to facilitate the discussion, a few words may be committed to the notation involved. A uniform symbol $C^{b}_{order_s, disorder_s}$ is used; where the two subscripts refer to two order (E, I) and disorder (S, R) parameters. Another subscript s is used to signify the space; *viz*., r, p or t (total). Two scaling parameters $b = \frac{2}{3}, 1$ are illustrated with superscripts (1), (2). In this part we will discuss the outcome obtained from a recent complexity analysis undertaken by us [8].

In CHA, it can be established that, out of these eight measures, $C_{ES}^{(2)}, E_{ER}^{(2)}$ and $C_{IS}^{(1)}, E_{IR}^{(1)}$ can detect the presence of node in a given n, l, m state. Because, in all these four measures, there appear extrema (both in r, p spaces) in nodal states. However, in $C_{ES}^{(1)}, E_{ER}^{(1)}$ and $C_{IS}^{(2)}, E_{IR}^{(2)}$ no such extrema is observed. Thus, it can be said that former two pairs explain CHA better in comparison to the latter two pairs. Hence, depending upon the nature of complexity measures, it is required to determine a proper value of b.

4.2.2. Relative Information for Hydrogen Atom

The radial wave function in r space simplifies to commonly used form, as given below,

$$\psi_{n,l}(r) = \frac{2}{n^2}\left[\frac{(n-l-1)!}{(n+l)!}\right]^{\frac{1}{2}}\left[\frac{2Z}{n}r\right]^l e^{-\frac{Z}{n}r}\, L_{(n-l-1)}^{(2l+1)}\left(\frac{2Z}{n}r\right). \quad (59)$$

$L_{(n-l-1)}^{(2l+1)}(s)$ is associated Laguerre polynomial. Putting $\xi = \frac{2Zr}{n}$ in Eq. (56), one gets,

$$IR_{\mathbf{r}} = 4\left(\frac{n}{2Z}\right)\int_0^\infty \xi^2 R_{n,l}^2(\xi)\left[\frac{R'_{n,l}(\xi)}{R_{n,l}(\xi)} - \frac{R'_{n_1,l}(\xi)}{R_{n_1,l}(\xi)}\right]^2 \mathrm{d}\xi. \quad (60)$$

Where primes denote 1st-order derivatives with respect to ξ. Then one can write,

$$\frac{R'_{n,l}(\xi)}{R_{n,l}(\xi)} = \frac{l}{\xi} - \frac{1}{2} - \frac{L_{(n-l-2)}^{(2l+2)}(\xi)}{L_{(n-l-1)}^{(2l+1)}(\xi)}; \qquad \frac{R'_{n_1,l}(\xi)}{R_{n_1,l}(\xi)} = \frac{l}{\xi} - \frac{1}{2} - \frac{L_{(n_1-l-2)}^{(2l+2)}(\xi)}{L_{(n_1-l-1)}^{(2l+1)}(\xi)}, \quad (61)$$

since $R_{n_1,l}(\xi)$ represents a circular state. So, $L_{(n_1-l-1)}^{(2l+1)}(\xi)$ provides a constant term, and hence $L_{(n_1-l-2)}^{(2l+2)}(\xi) = 0$. This facilitate the second ratio as, $\frac{R'_{n_1,l}(\xi)}{R_{n_1,l}(\xi)} = \frac{l}{\xi} - \frac{1}{2}$. Using above criteria and orthonormality condition of Associated Laguerre polynomial, Eq. (60) gives,

$$IR_{\mathbf{r}} = \left(\frac{8}{Zn^3}\right)\frac{(n-l-1)!}{(n+l)!}\int_0^{\infty}\xi^{2l+2}e^{-\xi}\left[L_{(n-l-2)}^{(2l+2)}(\xi)\right]^2 \mathrm{d}\xi, \tag{62}$$

which, after some algebraic manipulation, gives the simplified form of $IR_{\mathbf{r}}$ as below,

$$IR_{\mathbf{r}} = \frac{8(n-l-1)}{Zn^3} \qquad (\text{when } n > l,\ n-l \geq 2). \tag{63}$$

Equation (63) clearly suggests that, $IR_{\mathbf{r}}$ reduces with rise of $n,\ l$ and Z. Thus with progress in $n,\ l$, spatial separation between distribution of reference circular and respective l state decreases. In other words, fluctuation of a particular state with respect to node-less reference state reduces with increment of nodes. It may be recalled that the behavioral pattern of $IR_{\mathbf{r}}$ with n is similar to that of $I_{\mathbf{r}}$ [81]–both diminish as n advances. However, $I_{\mathbf{r}}$ is independent of l, whereas, $IR_{\mathbf{r}}$ seems to reduce with growth of l, for a fixed n. Table 6 presents some representative $IR_{\mathbf{r}}$ results for $n_s s, n_p p, n_d d, n_f f (n_s \geq 2, n_p \geq 3, n_d \geq 4, n_f \geq 5)$ orbitals for H atom, considering the respective circular states as reference.

In case of *even-l* states maximum in $IR_{\mathbf{r}}$ appears at $n = \frac{(3l+4)}{2}$, the resembling value being $\frac{32}{(3l+4)^3}(l+2)$. For *odd-l*, the same occurs for $n = \frac{3}{2}(l+1)$, with a respective value of $\frac{32}{27}\frac{1}{(l+1)^2}$. The panels A(a)-A(b) of Fig. 1 (of ref. [12]) illustrate variations of $IR_{\mathbf{r}}$ with changes in n for *even* ($l = 2,\ 4,\ 6,\ 8$) and *odd* ($l = 3,\ 5,\ 7,\ 9$) l states respectively. Each curve passes through a maximum, which tends to shift towards right as l attains higher values. One also notices that when $n \gg l$ then, we achieve $IR_{\mathbf{r}} \approx -\frac{16}{Z^3}\mathcal{E}_n$.

Next, we move on to $IR_{\mathbf{p}}$. The analytical expression [214] for wave function reads,

$$\psi_{n,l}(p) = n^2\left[\frac{2}{\pi}\frac{(n-l-1)!}{(n+l)!}\right]^{\frac{1}{2}} 2^{(2l+2)}\, l!\, \frac{n^l}{\{[\frac{np}{Z}]^2+1\}^{l+2}}\left(\frac{p}{Z}\right)^l C_{n-l-1}^{l+1}\left(\frac{[\frac{np}{Z}]^2-1}{[\frac{np}{Z}]^2+1}\right), \tag{64}$$

where $C_{\zeta}^{\eta}(X)$ signifies the Gegenbauer polynomial. Let us consider $t = \frac{np}{Z}$, and subsequently $q = \frac{t^2-1}{t^2+1}$. These two substitutions transform Eq. (56) into the form,

$$IR_{\mathbf{p}} = Zl!^2n^3\left[\frac{2^{4l+7}}{\pi}\frac{(n-l-1)!}{(n+l)!}\right] *$$
$$* \int_{-1}^{1}(1+q)^{\frac{3}{2}}(1-q)^{\frac{5}{2}}[f_{n,l}(q)]^2\left[\frac{d}{dq}\left(\frac{f'_{n,l}(q)}{f_{n,l}(q)} - \frac{f'_{n_1,l}(q)}{f_{n_1,l}(q)}\right)\right]^2 dq, \tag{65}$$

Table 6. Some specimen $IR_{\mathbf{r}}$ results for $n_s s, n_p p, n_d d, n_f f$ ($n_s \geq 2, n_p \geq 3, n_d \geq 4, n_f \geq 5$) orbitals of H atom, considering the corresponding circular states as reference [12]. For $2s, 3s, 4s, 5s$ states, data given in parentheses represent literature results [213]

Orbital	$IR_{\mathbf{r}}$	Orbital	$IR_{\mathbf{r}}$	Orbital	$IR_{\mathbf{r}}$	Orbital	$IR_{\mathbf{r}}$
$2s$	1 (1)	$3p$	$\frac{8}{27}$	$4d$	$\frac{1}{8}$	$5f$	$\frac{8}{125}$
$3s$	$\frac{16}{27}$ (0.5925)	$4p$	$\frac{1}{4}$	$5d$	$\frac{16}{125}$	$6f$	$\frac{2}{27}$
$4s$	$\frac{3}{8}$ (0.375)	$5p$	$\frac{24}{125}$	$6d$	$\frac{1}{9}$	$7f$	$\frac{24}{343}$
$5s$	$\frac{32}{125}$ (0.256)	$6p$	$\frac{4}{27}$	$7d$	$\frac{32}{343}$	$8f$	$\frac{1}{6}$

where $f'_{k,l}(q) = \frac{\mathrm{d}}{\mathrm{d}q}[f_{k,l}(q)]$, $f_{(k,l)}(q) = (1+q)^{\frac{l}{2}}(1-q)^{\frac{l}{2}+2}C_{k-l-1}^{l+1}(q)$. Also we note that,

$$\begin{aligned}
\frac{f'_{n,l}(q)}{f_{n,l}(q)} &= \frac{l}{2(1+q)} - \frac{\frac{l}{2}+2}{1-q} + \frac{\left(C_{n-l-1}^{l+1}(q)\right)'}{C_{n-l-1}^{l+1}(q)} \\
\frac{f'_{n_1,l}(q)}{f_{n_1,l}(q)} &= \frac{l}{2(1+q)} - \frac{\frac{l}{2}+2}{1-q} + \frac{\left(C_{n_1-l-1}^{l+1}(q)\right)'}{C_{n_1-l-1}^{l+1}(q)}.
\end{aligned} \tag{66}$$

Here too, as in case of $L_n^{\alpha}(x)$ for $IR_{\mathbf{r}}$, $C_{n_1-l-1}^{l+1}(q)$, being a part of circular state, is a constant and hence $(C_{n_1-l-1}^{l+1}(q))' = 0$. Therefore, one may write,

$$\frac{f'_{n_1,l}(q)}{f_{n_1,l}(q)} = \frac{l}{2(1+q)} - \frac{\frac{l}{2}+2}{1-q}. \tag{67}$$

After going through some algebra, one gets the following expression,

$$IR_{\mathbf{p}} = Z\, l!^2\, n^3 \left[\frac{2^{4l+7}}{\pi}\, \frac{(n-l-1)!}{(n+l)!}\right] (l+1)^2\, (I_1 + I_2)\,, \tag{68}$$

where the two integrations are defined as,

$$I_1 = \int_{-1}^{1} (1-q^2)^{l+\frac{3}{2}} \left[C_{n-l-2}^{l+2}(q)\right]^2 \mathrm{d}q,\; I_2 = \int_{-1}^{1} q(1-q^2)^{l+\frac{3}{2}} \left[C_{n-l-2}^{l+2}(q)\right]^2 \mathrm{d}q \tag{69}$$

Using the fact that, $I_2 = 0$ as the integrand is an odd function of q, finally we obtain,

$$\begin{aligned} IR_{\mathbf{p}} &= Z\, l!^2\, n^3 \left[\frac{2^{4l+7}}{\pi}\frac{(n-l-1)!}{(n+l)!}\right](l+1)^2 I_1 \\ &= 2^4\, Z\, n^2[n^2-(l+1)^2] \quad (\text{when}\, n>l,\; n-l\geq 2). \end{aligned} \tag{70}$$

This equation suggests that, at a particular l, $IR_{\mathbf{p}}$ enhances with n. On the contrary, at a certain n, like $IR_{\mathbf{r}}$, it decreases with l. In panels B(a), B(b) of Fig. 1 of ref. [12], $\ln(IR_{\mathbf{p}})$ is plotted against n for the same set of even and odd-l states (discussed in $IR_{\mathbf{r}}$ part) respectively. These two graphs conclude that, there is neither a maximum nor a minimum in $IR_{\mathbf{p}}$. In this case, when $n \gg l, IR_{\mathbf{p}} \approx \frac{4Z^5}{\mathcal{E}_n^2}$.

$IR_{\mathbf{r}}$ measures the change of fluctuation from ground (fixed l) to excited states. Thus, in a nutshell it can be inferred that, result of $IR_{\mathbf{r}}$ in H atom provides more detailed information than I, as it incorporates the effect of l in it [81]. Not only that, it also reinforce the epilogue that, the diffused nature of an orbital with constant l increases with increment of n.

4.2.3. A Virial-like Theorem for CHA

Recently a virial-like equation has been designed by the authors for confined quantum systems [215], which takes the following form,

$$\begin{aligned} \langle \hat{T}^2\rangle_n - \langle \hat{T}\rangle_n^2 &= \langle \hat{V}^2\rangle_n - \langle \hat{V}\rangle_n^2 \\ (\Delta\hat{T}_n)^2 = \langle \hat{V}\rangle_n\langle \hat{T}\rangle_n - \langle \hat{T}\hat{V}\rangle_n &= (\Delta\hat{V}_n)^2 = \langle \hat{T}\rangle_n\langle \hat{V}\rangle_n - \langle \hat{V}\hat{T}\rangle_n. \end{aligned} \tag{71}$$

This relation concludes that, the magnitude of error associated with $\langle \hat{T}\rangle_n$ and $\langle \hat{V}\rangle_n$ are identical. Now, one can easily infer that, $\mathcal{E}_n$ is a sum of two average quantities but still provides exact result. It is because of the cancellation of errors between $\langle \hat{T}\rangle_n$ and $\langle \hat{V}\rangle_n$. In a confinement condition, validity of Eq. (71) can be checked by deriving the expressions of $\langle \hat{T}\hat{V}\rangle_{n,\ell}$, $\langle \hat{V}\hat{T}\rangle_{n,\ell}$, $\langle \hat{V}^2\rangle_{n,\ell}$ and $\langle \hat{V}\rangle_{n,\ell}$ (other integrals remain unchanged).

We begin with the *exact* wave function for CHA given in Eq. (21). The pertinent expectation values can be simplified as,

$$\langle \hat{T}\hat{V}\rangle_{n,\ell} = \langle \hat{T}v(r)\rangle_{n,\ell} + \langle \hat{T}v_c(r)\rangle_{n,\ell} = \langle \hat{T}v(r)\rangle_{n,\ell}. \tag{72}$$

Table 7. $\mathcal{E}_{n,\ell}$, $(\Delta V_{n,\ell})^2$, $(\Delta T_{n,\ell})^2$, $\langle T\rangle_{n,\ell}\langle V\rangle_{n,\ell} - \langle TV\rangle_{n,\ell}$, $\langle T\rangle_{n,\ell}\langle V\rangle_{n,\ell} - \langle VT\rangle_{n,\ell}$ of $1s, 2s, 2p$ states in CHA at four $(0.1, 0.2, 0.5, \infty)$ r_c's [215]. See text for detail

State	Property	$r_c = 0.1$	$r_c = 0.2$	$r_c = 0.5$	$r_c = \infty$
	$\mathcal{E}_{1,0}^{\P}$	468.9930386	111.0698588	14.74797003	−0.499999
	$(\Delta V_{1,0})^2$	308.8728899	80.38083598	14.53962018	0.99999999
$1s$	$(\Delta T_{1,0})^2$	308.8728899	80.38083598	14.53962018	0.99999999
	$\langle T\rangle_{1,0}\langle V\rangle_{1,0} - \langle TV\rangle_{1,0}$	308.8728899	80.38083598	14.53962018	0.99999999
	$\langle T\rangle_{1,0}\langle V\rangle_{1,0} - \langle VT\rangle_{1,0}$	308.8728899	80.38083598	14.53962018	0.99999999
	$\mathcal{E}_{n,l}^{\dagger}$	1942.720354	477.8516723	72.67203919	−0.124999
	$(\Delta V_{2,0})^2$	925.8428960	236.7351455	40.51345969	0.18749999
$2s$	$(\Delta T_{2,0})^2$	925.8428960	236.7351455	40.51345969	0.18749999
	$\langle T\rangle_{2,0}\langle V\rangle_{2,0} - \langle TV\rangle_{2,0}$	925.8428960	236.7351455	40.51345969	0.18749999
	$\langle T\rangle_{2,0}\langle V\rangle_{2,0} - \langle VT\rangle_{2,0}$	925.8428960	236.7351455	40.51345969	0.18749999
	$\mathcal{E}_{2,1}^{\ddagger}$	991.0075894	243.1093321	36.65887588	−0.124999
	$(\Delta V_{2,1})^2$	47.98046148	12.14249373	2.016203448	0.02083333
$2p$	$(\Delta T_{2,1})^2$	47.98046148	12.14249373	2.016203448	0.02083333
	$\langle T\rangle_{2,1}\langle V\rangle_{2,1} - \langle TV\rangle_{2,1}$	47.98046148	12.14249373	2.016203448	0.02083333
	$\langle T\rangle_{2,1}\langle V\rangle_{2,1} - \langle VT\rangle_{2,1}$	47.98046148	12.14249373	2.016203448	0.02083333

¶Literature results [4] of $\mathcal{E}_{1,0}$ for $r_c = 0.1, 0.2, 0.5, \infty$ are: 468.9930386595, 111.0698588367, 14.74797003035, −0.5 respectively.

†Literature results [4] of $\mathcal{E}_{2,0}$ for $r_c = 0.1, 0.2, 0.5, \infty$ are: 1942.720354554, 477.8516723922, 72.67203919047, −0.125 respectively.

‡ Literature results [4] of $\mathcal{E}_{2,1}$ for $r_c = 0.1, 0.2, 0.5, \infty$ are: 991.0075894412, 243.1093166600, 36.65887588018, −0.125 respectively.

In this instance, $\langle \hat{T} v_c(r)\rangle_{n,\ell} = 0$, as the wave function disappears for $r \geq r_c$. Use of same argument, along with the fact that $\langle v_c(r)\hat{T}\rangle_{n,\ell} = 0$, gives rise to,

$$\langle \hat{V}\hat{T}\rangle_{n,\ell} = \langle v(r)\hat{T}\rangle_{n,\ell} + \langle v_c(r)\hat{T}\rangle_{n,\ell} = \langle v(r)\hat{T}\rangle_{n,\ell}. \tag{73}$$

Now, since $\langle v(r)v_c(r)\rangle_{n,\ell} = \langle v_c(r)v(r)\rangle_{n,\ell} = \langle v_c(r)^2\rangle_{n,\ell} = 0$, one may write,

$$\langle \hat{V}^2\rangle_{n,\ell} = \langle v(r)^2\rangle_{n,\ell} + \langle v(r)v_c(r)\rangle_{n,\ell} + \langle v_c(r)v(r)\rangle_{n,\ell} + \langle v_c(r)^2\rangle_{n,\ell} = \langle v(r)^2\rangle_{n,\ell}. \tag{74}$$

Again, because $\langle v_c(r)\rangle_{n,\ell} = 0$, it follows that,

$$\langle \hat{V}\rangle_{n,\ell} = \langle v(r)\rangle_{n,\ell} + \langle v_c(r)\rangle_{n,\ell} = \langle v(r)\rangle_{n,\ell}. \tag{75}$$

Thus, for a CHA also, Eq. (71) remains unchanged, i.e.,

$$\langle \hat{T}^2 \rangle_{n,\ell} - \langle \hat{T} \rangle^2_{n,\ell} = \langle \hat{V}^2 \rangle_{n,\ell} - \langle \hat{V} \rangle^2_{n,\ell}$$

$$\left(\Delta \hat{T}_{n,\ell}\right)^2 = \left(\Delta \hat{V}_{n,\ell}\right)^2 = \langle \hat{T} \rangle_{n,\ell} \langle v(r) \rangle_{n,\ell} - \langle v(r) \hat{T} \rangle_{n,\ell} = \langle \hat{T} \rangle_{n,\ell} \langle v(r) \rangle_{n,\ell} - \langle \hat{T} v(r) \rangle_{n,\ell}. \tag{76}$$

This equation suggests that, CHA fulfills the results given in Eq. (71); and v_c has no impact on it. It has only introduced the boundary in a finite range. Table 7 demonstrates sample values of $\mathcal{E}_{n,\ell}$, $(\Delta \hat{T}_{n,\ell})^2$, $(\Delta \hat{V}_{n,\ell})^2$, $\langle T \rangle_{n,\ell} \langle V \rangle_{n,\ell} - \langle TV \rangle_{n,\ell}$ and $\langle T \rangle_{n,\ell} \langle V \rangle_{n,\ell} - \langle VT \rangle_{n,\ell}$ for same low-lying ($1s$, $2s$, $2p$) states, in CHA at four particular r_c's, like 0.1, 0.2, 0.5, ∞. For sake of completeness, accurate values of $\mathcal{E}_{n,\ell}$ are reproduced from [4]. In both confining and free (last column) conditions, these results complement the conclusion of Eq. (71). In passing, it is interesting to note that both $(\Delta \hat{T}_{n,\ell})^2$, $(\Delta \hat{V}_{n,\ell})^2$ decrease with rise in r_c.

4.3. Many-electron Atom in Spherical Cage

In this subsection, we briefly cover some fundamental aspects regarding the application of DFT to many-electron atoms, where the electrons remain under the influence of an external potential that imposes restriction of an impenetrable spherical cavity. DFT has been successfully implemented in explaining the electronic structure and properties of many-electron systems in their ground state. It is, however, noted in the derivation of Hohenberg-Kohn theorems that, DFT is usually termed as a ground-state theory as the practical approach to extract the information on excited-state energies and densities has remained a bottleneck. About a decade ago, an exchange potential based on the work-function approach [149] was proposed to perform both ground and excited state calculations in many-electron atoms. This non-variational method is simple, computationally efficient, and has shown considerable success for some singly, doubly, and triply excited states, low and moderately high states, valence and core excitations, as well as in explaining the auto-ionizing and satellite states for free systems [216, 217, 218, 219, 220, 143, 146, 147, 148]. However, while the success of this method is well-established and well documented, it has never been tested so far, in the context of confinement studies, such as those undertaken here. Therefore, it will be a very much worthwhile investigation. Due to the existence of Coulomb singularity at the origin and its long-range nature, we employ the GPS procedure to achieve high accuracy. It allows nonuniform spatial grid discretization and considerably more accurate calculations of eigenvalues and wave functions can be obtained using only a modest number of grid points. Here we shall present some new complementary results for further evaluation of the methodology in regards to *hard* confinement.

4.3.1. Spherically Confined Atoms

In order to see the contribution of each part of the exchange-correlation (XC) functionals, in Table 8 we present the X-only energy of confined He atom in its ground state obtained by work-function approximation as a function of confining radius and they are contrasted with the HF results reported in [221]. The second and third columns of this table stand for the X-only results calculated with our method and HF results respectively. Also total energy considering the local and gradient corrected correlation functionals are reported in fourth and fifth columns. Next two columns quote energies reported in the literature obtained with wave function based-approaches; the multi-configuration parametrized optimized effective potential (JPOEP) method [62] and variational energies calculated with Hylleraas wave function [222]. From this table the appreciable conclusions which can be drawn are: i) results obtained from our proposed method indicate the correct behavior of total energy as a function of confinement radii when compared with corresponding available information for He, and ii) our X-only results are in good agreement with HF values. For all r_c's the X-only energies match up to four decimal points with HF energies. Predictably, the local correlation functionals (here Wigner) perform worse than the gradient-corrected (LYP) one. Considering Hylleraas wave function to be the best variational wave function for two-electron systems, we may conclude that LYP functional works well for weak confinements but suffers in small r_c region, which needs to be addressed in future.

We have particular interest in assessing the accuracy and reliability to describe the excited states of confined many-electron atom, for which literature results are quite scanty. In Table 9, the interplay between correlations and confinement is analyzed for first excited ($1s2s\ ^3S$) state of He. These energies were calculated over a range of r_c varying from 0.1 to 10.0 (in a.u.). Selected energies from this work, with Wigner and LYP correlation functionals, and those calculated with Hylleraas wave function from [221] are recorded in second, third and fourth columns respectively. It is clear that both methods predict a marked increase in energy, as we reduce the confinement radius. An analysis

Table 8. Ground-state energy of confined He at some chosen r_c. See text for details

r_c	E_X	E_{HF}	E_{XC}(Wigner)	E_{XC}(LYP)	E(JPOEP)	E(Hylleraas)
0.1	906.61645	—	906.4496	907.455	—	906.562423
0.5	22.79096	22.79095	22.6909	22.9592	—	22.741303
0.8	4.65737	4.65739	4.5787	4.7289	—	—
1.0	1.06121	1.06122	0.9915	1.0988	1.0158	1.015755
2.0	−2.56256	−2.56251	−2.6115	−2.5879	−2.6040	−2.604038
3.0	−2.83103	−2.83102	−2.8746	−2.8688	−2.8724	−2.872494
5.0	−2.86136	−2.86129	−2.9035	−2.9033	−2.9033	−2.903410
100.0	−2.86164	−2.86164	−2.9037	−2.9053	−2.9036	−2.906956

Table 9. Single excitation (1s2s, ^3S) energy in confined He at some chosen r_c

r_c	E_X	E_{XC}(Wigner)	E_{XC}(LYP)	E(Hylleraas)
0.1	2370.7389	2370.5729	2372.1569	2370.7270
0.5	78.83157	78.7330	79.1755	78.8208
1.0	14.36896	14.3014	14.4954	14.3597
2.0	0.56698	0.5227	0.5795	0.5602
3.0	−1.36562	−1.4009	−1.3822	−1.3705
5.0	−2.04515	−2.0740	−2.0732	−2.0480
10.0	−2.17079	−2.1964	−2.1985	−2.17262

of XC potential for this particular case shows that, both the correlation energy functionals can provide reasonable good agreements with other sophisticated wave-function-based approaches (considering energies for Hylleraas to be the most authentic one).

Conclusion

In this chapter, at first we have summarized the role of I, S, E and Onicescu-Shannon complexity in understanding the quasi-degeneracy and trapping of a particle within either of the wells in an SDW potential. Later, we employed the same information measures to explore oscillatory nature of a particle in an ADW potential. A thorough analysis exposes that, in SDW potential, conventional uncertainty measures are unable to explain the various rich facets (such as quasi-degeneracy, localization of particle) associated with it, in a satisfactory manner. However, investigation of S, E suggests that, there occurs an interplay between localization and delocalization. At first delocalization predominates with increase in S_x and decrease in E_x, but after a characteristic β, this trend reverses–localization prevails. It is imperative to mention that, in this context, S_x passes through a maximum and E_x comes out from a minimum. In case of ADW potential, a pair of rules have been demonstrated to discern quasi-degeneracy and to-and-fro motion of a particle between deeper and shallower wells. It also suggests that (i) a particle in an nth state oscillates n times before localizing in deeper well (ii) at certain sets of α, β, γ, ADW behaves as two different potentials.

Next, the effect of moving the boundary from infinity to finite region has been delineated in CHA through numerous information quantities quite decently uncovering many hitherto unreported interesting features. A systematic and elaborate survey leads to the fact that, at sufficiently low r_c, the patterns in I, S, R, E get reversed to that encountered in FHA. This interprets that, the effect of confinement is more on higher quantum states. Apart from that, the feasibility and accessibility of a new virial-like theorem, recently proposed by us, was tested in this context. As expected, this turns out to be quite successful and physically meaningful. Further, relative Fisher information for FHA are also reported here. Lastly, we inspected the influence of similar boundary condition on ground and excited states of helium atom using a work-function based recipe within DFT. This proposed new method offers correct physical behavior in confinement in both ground and excited states. The illustrative results presented here for the first time, demonstrate its promise, and may be pursued for other many-electron atoms, in future. Two representative correlation energy functionals were put to test to account for the subtle effects of electron correlation. While they produce better energies, from exchange-only calculations, there is clearly a need for more accurate and reliable functionals, especially in stronger confinement region. On the other side, however, it is abundantly clear that, this single-determinantal approach offers very competitive excited-state energies compared to other wave function-based multi-configuration or basis-set-related sophisticated methods. In this case, only energy analysis has been provided. Information analysis for confined helium as well as other atoms will be undertaken in due course. In future, it will be desirable to explore the effect of more realistic (penetrable) boundary on such systems. Further, investigation of relative information in confined atoms using various reference states would be highly useful.

Acknowledgments

The authors sincerely thank Ms. Nadya S. Columbus, President, NOVA Science Publishers, NY, USA, and the Editor Dr. Vijay Kumar, for the kind invitation to contribute a chapter in this topical area. AKR gratefully acknowledges financial support from SERB INDIA, New Delhi (sanction no. CRG/2019/000293, MTR/2019/000012) and CSIR INDIA (sanction no. 01/(3027)/21/EMR-II).

References

[1] Sen, K. D.; Roy, A. K. Studies on the 3D confined potentials using generalized pseudospectral approach. *Phys. Lett. A*, 357:112, 2006.

[2] Roy, A. K. Confinement in 3D polynomial oscillators through a generalized pseudospectral method. *Mod. Phys. Lett. A*, 29:1450104, 2014.

[3] Roy, A. K. Studies on bound-state spectra of Manning-Rosen potential. *Mod. Phys. Lett. A*, 29:1450042, 2014.

[4] Roy, A. K. Spherical confinement of Coulombic systems inside an impenetrable box: H atom and the Húlthen potential. *Int. J. Quant. Chem.*, 115:937, 2015.

[5] Roy, A. K. Critical parameters and spherical confinement of H atoms in screened Coulomb potential. *Int. J. Quant. Chem.*, 116:953, 2016.

[6] Mukherjee, N.; Roy, A.; Roy, A. K. Information entropy as a measure of tunneling and quantum confinement in a symmetric double-well potential. *Ann. Phys. (Berlin)*, 527:825, 2015.

[7] Mukherjee, N.; Roy, A. K. Quantum confinement in an asymmetric double-well potential through energy analysis and information-entropic measure. *Ann. Phys. (Berlin)*, 528:412, 2016.

[8] Majumdar, S.; Mukherjee, N.; Roy, A. K. Various complexity measures in confined hydrogen atom. *Chem. Phys. Lett.*, 687:322, 2017.

[9] Mukherjee, N.; Roy, A. K. Fisher information in confined isotropic harmonic oscillator. *Int. J. Quant. Chem.*, 118:e25727, 2018.

[10] Mukherjee, N.; Roy, A. K. Information−entropic measures in confined isotropic harmonic oscillator. *Adv. Theory Simul.*, 1:1800090, 2018.

[11] Mukherjee, N; Roy, A. K. Information-entropic measures in free and confined hydrogen atom. *Int. J. Quant. Chem.*, 118:e25596, 2018.

[12] Mukherjee, N.; Roy, A. K. Relative Fisher information in some central potentials. *Ann. Phys. (N.Y.)*, 398:190, 2018.

[13] Mukherjee, N.; Majumdar, S.; Roy, A. K. Fisher information in confined hydrogen-like ions. *Chem. Phys. Lett.*, 691:449, 2018.

[14] Majumdar, S.; Roy, A. K. Shannon entropy in confined he-like ions within a density functional formalism. *Quant. Rep.*, 2:189, 2020.

[15] Mukherjee, N.; Roy, A. K. Shell-confined atom and plasma: incidental degeneracy, metallic character, and information entropy. *Phys. Rev. A*, 104:042803, 2021.

[16] Majumdar, S.; Roy, A. K. Density functional study of atoms spatially confined inside a hard sphere. *Int. J. Quant. Chem.*, 121:026630, 2021.

[17] Majumdar, S.; Mukherjee, N.; Roy, A. K. Confined H^- ion within a density functional framework. *Eur. Phys. J. D*, 75:1, 2021.

[18] Jaskólski, W. Confined many-electron systems. *Phys. Rep.*, 271:1, 1996.

[19] Sabin, J.; Brändas, E.; and Cruz, S., editor. *Adv. Quant. Chem.*, volume 57 & 58. Academic Press, New York, 2009.

[20] Sen, K. D., editor. *Electronic Structure of Quantum Confined Atoms and Molecules.* Springer International Publishing, Switzerland, 2014.

[21] Dolmatov, V. K.; Baltenkov, A. S.; Connerade, J.-P; Mason, S. T. Structure and photoionization of confined atoms. *Radiat. Phys. Chem.*, 70:417, 2004.

[22] Grochala, W.; Hoffman, R.; Feng, J.; Ashcroft, N. W. The chemical imagination at work in very tight places. *Ang. Chem. Int. Ed.*, 46:3620, 2007.

[23] Michels, A.; De Boer, J.; Bijl, A. Remarks concerning molecular interaction and their influence on the polarizability. *Physica*, 4:981, 1937.

[24] De Groot, S. R.; Ten Seldam, C. A. On the energy levels of a model of the compressed hydrogen atom. *Physica*, 12:669, 1946.

[25] Ley-Koo, E.; Rubinstein, S. The hydrogen atom within spherical boxes with penetrable walls. *J. Chem. Phys.*, 71:351, 1979.

[26] Ley-Koo, E.; Rubinstein, S. The hydrogen atom inside boxes with paraboloidal surfaces. *J. Chem. Phys.*, 73:887, 1980.

[27] Marin, J. L.; Cruz, S. A. On the use of direct variational methods to study confined quantum systems. *Am. J. Phys.*, 59:931, 1991.

[28] Marin, J. L.; Cruz, S. A. Use of the direct variational method for the study of one- and two-electron atomic systems confined by spherical penetrable boxes. *J. Phys. B*, 25:4365, 1992.

[29] Aquino, N. Accurate energy eigenvalues for enclosed hydrogen atom within spherical impenetrable boxes. *Int. J. Quant. Chem.*, 54:107, 1995.

[30] N. Aquino. The hydrogen and helium atoms confined in spherical boxes. In *Advances in Quantum Chemistry*, chapter 4, pages 123–171. Elsevier Publishers, Hauppauge, New York, 2009.

[31] Katriel, J.; Montgomery Jr., H. E. The virial theorem for the smoothly and sharply, penetrably and impenetrably confined hydrogen atom. *J. Chem. Phys.*, 137:114109, 2012.

[32] Le Sech, C.; Banerjee, A. A variational approach to the Dirichlet boundary condition: Application to confined H^-, He and Li. *J. Phys. B.*, 44:105003, 2011.

[33] Sarsa, A.; Le Sech, C. Variational Monte Carlo method with Dirichlet boundary conditions: Application to the study of confined systems by impenetrable surfaces with different symmetries. *J. Chem. Thoery. Comput.*, 7:2786, 2011.

[34] Cabrera-trujillo, R.; Cruz, S. A. Confinement approach to pressure effects on the dipole and the generalized oscillator strength of atomic hydrogen. *Phys. Rev. A*, 87:012502, 2013.

[35] Connerade, J. P.; Dolmatov, V. K.; Lakshmi, P.A. The filling of shells in compressed atoms. *J. Phys. B*, 33:251, 2000.

[36] Colín-Rodríguez, R.; Díaz-García, C.; Cruz, S. A. The hydrogen molecule and the H_2^+ molecular ion inside padded prolate spheroidal cavities with arbitrary nuclear positions. *J. Phys. B*, 44:241001, 2011.

[37] Montgomery Jr., H. E.; Sen, K. D. Dipole polarizabilities for a hydrogen atom confined in a penetrable sphere. *Phys. Lett. A*, 376:1992, 2012.

[38] Wigner, E.; Seitz, F. On the constitution of metallic sodium. *Phys. Rev.* , 43:804, 1933.

[39] Wigner, E.; Seitz, F. On the constitution of metallic sodium II. *Phys. Rev.* , 46:509, 1934.

[40] Sobrino-Coll, N.; Puertas-Centeno, D.; Toranzo, I. V.; Dehesa, J. S. Complexity measures and uncertainty relations of the high-dimensional harmonic and hydrogenic systems. *J. Stat. Mech.*, 8:083102, 2017.

[41] Goldman, S.; Joslin, C. Spectroscopic properties of an isotropically compressed hydrogen atom. *J. Phys. Chem.*, 96:6021, 1992.

[42] Garza, J.; Vargas, R.; Vela, A. Numerical self-consistent-field method to solve the Kohn-Sham equations in confined many-electron atoms. *Phys. Rev. E*, 58:3949, 1998.

[43] Laughlin, C.; Burrows, B. L.; Cohen, M. A hydrogen-like atom confined within an impenetrable spherical box. *J. Phys. B*, 35:701, 2002.

[44] Burrows, B. L.; Cohen, M. Exact solutions for spherically confined hydrogen-like atoms. *Int. J. Quant. Chem.*, 106:478, 2006.

[45] Aquino, N.; Campoy, G.; Montgomery Jr., H. E. Highly accurate solutions for the confined hydrogen atom. *Int. J. Quant. Chem.*, 107:1548, 2007.

[46] Baye, D.; Sen, K. D. Confined hydrogen atom by the Lagrange-mesh method: Energies, mean radii, and dynamic polarizabilities. *Phys. Rev. E*, 78:026701, 2008.

[47] Ciftci, H.; Hall, R. L.; Saad, N. Study of a confined hydrogen-like atom by the asymptotic iteration method. *Int. J. Quant. Chem.*, 109:931, 2009.

[48] Sen, K. D. Characteristic features of Shannon information entropy of confined atoms. *J. Chem. Phys.*, 123:074110, 2005.

[49] Laughlin, C.; Chu, S. I. A highly accurate study of a helium atom under pressure. *J. Phys. A*, 42:265004, 2009.

[50] Gorecki, J.; Byers-Brown, W. Padded-box model for the effect of pressure on helium. *J. Phys. B*, 21:403, 1988.

[51] Rodriguez-Bautista, M.; Díaz-García, C.; Navarrete-López, A. M.; Vargas, R.; Garza, J. Roothaans approach to solve the Hartree-Fock equations for atoms confined by soft walls: Basis set with correct asymptotic behavior. *J. Chem. Phys.*, 143:034103, 2015.

[52] Sarsa, A.; Alcaraz-Pelegrina, J. M.; Le Sech, C. Isotopic effects on covalent bond confined in a penetrable sphere. *J. Phys. Chem.*, 119:14364, 2015.

[53] Ten Seldam, C. A.; De Groot, R. On the ground state of a model for compressed helium. *Physica*, 18:891, 1952.

[54] Aquino, N.; Flores-Riveros, A.; Rivas-Silva, J. F.; Sen, K. D. Confined helium atom low-lying S states analyzed through correlated Hylleraas wave functions and the Kohn-Sham model. *J. Chem. Phys.*, 124:054311, 2006.

[55] Wilson, C. L.; Montgomery Jr., H. E.; Sen, K. D.; Thompson, D. C. Electron correlation energy in confined two-electron systems. *Phys. Lett. A*, 374:4415, 2010.

[56] Flores-Riveros, A.; Aquino, N.; Montgomery Jr., H. E. Spherically compressed helium atom described by perturbative and variational methods. *Phys. Lett. A*, 374:1246, 2010.

[57] Flores-Riveros, A.; Rodríguez-Contreras, A. Compression effects in helium-like atoms ($Z = 1, 5$) constrained by hard spherical walls. *Phys. Lett. A*, 372:6175, 2008.

[58] Sakiroglu, S.; Dogan, Ü.; Yildiz, A.; Akgüngör, K.; Epik, H.; Ergün, Y.; Sari, H.; Sökmen, I. Ground-state energy of excitons in quantum dot treated variationally via Hylleraas-like wave function. *Chin. Phys. B*, 18:1578, 2009.

[59] Sarkar, U.; Giri, S.; Chattaraj, P. K. Dirichlet boundary conditions and effect of confinement on chemical reactivity. *J. Phys. Chem. A*, 113:10759, 2009.

[60] van Faassen, M. Atoms in boxes: From confined atoms to electron-atom scattering. *J. Chem. Phys.*, 131:104108, 2009.

[61] Sarsa, A.; Buendía, E.; Gálvez, F. J. Study of confined many-electron atoms by means of the POEP method. *J. Phys. B.*, 47:185002, 2014.

[62] Sarsa, A.; Buendía, E.; Gálvez, F. J. Multi-configurational explicitly correlated wave functions for the study of confined many-electron atoms. *J. Phys. B.*, 49:145003, 2016.

[63] Gálvez, F. J.; Buendía, E.; Sarsa, A. Confinement effects on the electronic structure of M-shell atoms: A study with explicitly correlated wave functions. *Int. J. Quant. Chem.*, 117:e25421, 2017.

[64] Gimarc, B. M. Correlation energy of the two-electron atom in a spherical potential box. *J. Chem. Phys.*, 47:5110, 1967.

[65] Rivelino, R.; Vianna, J. D. M. A configuration interaction model to investigate many-electron systems in cavities. *J. Phys. B*, 34:L645, 2001.

[66] Ludeña, E. V. SCF Hartree-Fock calculations of ground state wave functions of compressed atoms. *J. Chem. Phys.*, 69:1770, 1978.

[67] Ludeña, E. V.; Gregori, M. Configuration interaction calculations for two-electron atoms in a spherical box. *J. Chem. Phys.*, 71:2235, 1979.

[68] Joslin, C.; Goldman, S. Quantum Monte Carlo studies of two-electron atoms constrained in spherical boxes. *J. Phys. B*, 25:1965, 1992.

[69] Whitkop, P. G. Integral evaluation in the case of a helium atom positioned off-center in a spherical box. *Int. J. Quant. Chem.*, 73:459, 1999.

[70] Poland, D. Maximum-entropy calculation of energy distributions. *J. Chem. Phys.*, 112:6554, 2000.

[71] Singer, A. Maximum entropy formulation of the Kirkwood superposition approximation. *J. Chem. Phys.*, 121:3657, 2004.

[72] Antoniazzi, A.; Fanelli, D.; Barré, J.; Chavanis, P.-H.; Dauxois, T.; Ruffo, S. Maximum entropy principle explains quasi-stationary states in systems with long-range interactions: The example of the Hamiltonian mean-field model. *Phys. Rev. E*, 75:011112, 2007.

[73] Bialyniki-Birula, I.; Mycielski, J. Uncertainty relation for information entropy in wave mechanics. *Commun. Math. Phys.*, 44:129, 1975.

[74] C. E. Shannon. Prediction and entropy of printed english. *Bell Syst. Tech. J.*, 30:50, 1951.

[75] Bialyniki-Birula, I. Formulation of uncertainty relations in terms of the Rényi entropies. *Phys. Rev. A*, 74:052101, 2006.

[76] Duan, L.-M.; Giedke. G.; Cirac, J. I.; Zoller, P. Inseparability criterion for continuous variable systems. *Phys. Rev. Lett.*, 84:2722, 2000.

[77] R. Simon. Peres-Horodecki separability criterion for continuous variable systems. *Phys. Rev. Lett.*, 84:2726, 2000.

[78] González-Férez, R.; Dehesa, J. S. Shannon entropy as an indicator of atomic avoided crossings in strong parallel magnetic and electric fields. *Phys. Rev. Lett.*, 91:113001, 2003.

[79] Sen, K. D., editor. *Statistical Complexity: Application in Electronic Structure*. Springer International Publishing, Berlin, Germany, 2012.

[80] Verguilla-Berdecia, L. A. Tunneling in a quartic, symmetric, double well potential. *J. Chem. Edu.*, 70:928, 1993.

[81] Romera, E.; Sánchez-Moreno, P.; Dehesa, J. S. The Fisher information of single-particle systems with a central potential. *Chem. Phys. Lett.*, 414:468, 2005.

[82] Romera, E.; Sánchez-Moreno, P.; Dehesa, J. S. Uncertainty relation for Fisher information of D-dimensional single-particle systems with central potentials. *J. Math. Phys.*, 47:103504, 2006.

[83] Dehesa, J. S.; López-Rosa, S.; Olmos, B. Fisher information of D-dimensional hydrogenic systems in position and momentum spaces. *J. Math. Phys.*, 47:052104, 2006.

[84] Dehesa, J. S.; González-Férez; Sánchez-Moreno, P. The Fisher-information-based uncertainty relation, Cramer-Rao inequality and kinetic energy for the D-dimensional central problem. *J. Phys. A*, 40:1845, 2007.

[85] Sánchez-Moreno, P.; Zozor, S.; Dehesa, J. S. Upper bounds on Shannon and Rényi entropies for central potentials. *J. Math. Phys.*, 52:022105, 2011.

[86] Toranzo, I. V.; Puertas-Centeno, D.; Dehesa, J. S. Entropic properties of D-dimensional Rydberg systems. *Physica A*, 462:1197, 2016.

[87] Yáñez, R. J.; Van Assche, W.; Dehesa, J. S. Position and momentum information entropies of the D-dimensional harmonic oscillator and hydrogen atom. *Phys. Rev. A*, 50:3065, 1994.

[88] Toranzo, I. V.; Dehesa, J. S. Rényi, Shannon and Tsallis entropies of Rydberg hydrogenic systems. *Europhys. Lett.*, 113:48003, 2016.

[89] Katriel, J.; Sen, K. D. Relativistic effects on information measures for hydrogen-like atoms. *J. Comput. Appl. Math.*, 233:1399, 2010.

[90] Dehesa, J. S.; López-Rosa, S.; Martínez-Finkelshtein, A.; Yáñez, R. J. Information theory of D-dimensional hydrogenic systems: Application to circular and Rydberg states. *Int. J. Quant. Chem.*, 110:1529, 2010.

[91] Jiao, L. G.; Zan, L. R.; Zhang, Y. Z.; Ho, Y. K. Benchmark values of Shannon entropy for spherically confined hydrogen atom. *Int. J. Quant. Chem.*, 117:e25375, 2017.

[92] Patil, S. H.; Sen, K. D.; Watson, N. A.; Montgomery Jr., H. E. Characteristic features of net information measures for constrained Coulomb potentials. *J. Phys. B*, 40:2147, 2007.

[93] Aquino, N.; Flores-Riveros, A.; Rivas-Silva, J. F. Shannon and Fisher entropies for a hydrogen atom under soft spherical confinement. *Phys. Lett. A*, 377:2062, 2013.

[94] Mukherjee, N.; Roy, A. K. Information-entropic measures for non-zero l states of confined hydrogen-like ions. *Eur. Phys. J. D*, 72:118, 2018.

[95] Majumdar, S.; Mukherjee, N.; Roy, A. K. Information entropy and complexity measure in generalized Kratzer potential. *Chem. Phys. Lett.*, 716:257, 2019.

[96] Guevara, N. L.; Sagar, R. P.; Esquivel, R. O. Shannon-information entropy sum as a correlation measure in atomic systems. *Phys. Rev. A*, 67:012507, 2003.

[97] Guevara, N. L.; Sagar, R. P.; Esquivel, R. O. Local correlation measures in atomic systems. *J. Chem. Phys.*, 122:084101, 2005.

[98] Romera E., Nagy, Á. Rényi information of atoms. *Phys. Lett. A*, 372:4918, 2008.

[99] Nagy, Á.; Romera, E. Relative Rényi entropy for atoms. *Int. J. Quant. Chem.*, 109:2490, 2009.

[100] Antolín, J.; Angulo, J. C. Complexity analysis of ionization processes and iso-electronic series. *Int. J. Quant. Chem.*, 109:586, 2009.

[101] Antolín, J.; López-Rosa, S.; Angulo, J. C. Rényi complexities and information planes: Atomic structure in conjugated spaces. *Chem. Phys. Lett.*, 474:233, 2009.

[102] Zhou, X.-Y.; Rong, C.; Lu, T.; Zhou, P.; Liu, S. Information Functional Theory: Electronic Properties as Functionals of Information for Atoms and Molecules. *J. Phys. Chem. A*, 120:3634, 2016.

[103] Dehesa, J. S.; Toranzo, I. V.; Puertas-Centeno, D. Entropic measures of Rydberg-like harmonic states. *Int. J. Quant. Chem.*, 117:48, 2017.

[104] Nagy, Á. Shannon entropy density as a descriptor of Coulomb systems. *Chem. Phys. Lett.*, 556:355, 2013.

[105] Ou, J.-H.; Ho, Y. K. Benchmark calculations of Rényi, Tsallis entropies, and Onicescu information energy for ground state helium using correlated Hylleraas wave functions. *Int. J. Quant. Chem.*, 0:e25928, 2019.

[106] Sears, S. B.; Gadre, S. R. An information theoretic synthesis and analysis of Compton profiles. *J. Chem. Phys.*, 75:4626, 1981.

[107] Gadre, S. R. Information entropy and Thomas-Fermi theory. *Phys. Rev. A*, 30:620, 1984.

[108] Gadre, S. R.; Sears, S. B. Some novel characteristics of atomic information entropies. *Phys. Rev. A*, 32:2602, 1985.

[109] Dehesa, J. S.; Galvez, F. J.; Porras, I. Bounds to density-dependent quantities of D-dimensional many-particle systems in position and momentum spaces: Applications to atomic systems. *Phys. Rev. A*, 40:35, 1989.

[110] Tripathi, A. N.; Sagar, R. P.; Esquivel, R. O.; Smith Jr., V. H. Electron correlation in momentum space: The beryllium-atom isoelectronic sequence. *Phys. Rev. A*, 45:4385, 1992.

[111] Hô, M.; Sagar, R. P.; Smith Jr., V. H.; Esquivel, R. O. Atomic information entropies beyond the Hartree-Fock limit. *J. Phys. B*, 27:5149, 1994.

[112] Tripathi, A. N.; Smith Jr., V. H.; Sagar, R. P.; Esquivel, R. O. Electron correlation in momentum space for the neon-atom isoelectronic sequence from F^- through Ar^{8+}. *Phys. Rev. A*, 54:1877, 1996.

[113] Noorizadeh, S.; Shakerzadeh, E. Shannon entropy as a new measure of aromaticity, Shannon aromaticity. *Phys. Chem. Chem. Phys.*, 12:4742, 2010.

[114] Sagar, R. P.; Ramírez, J. C.; Esquivel, R. O.; Hó, M.; Smith Jr., V. H. Relationships between Jaynes entropy of the one-particle density matrix and Shannon entropy of the electron densities. *J. Chem. Phys.*, 116:9213, 2002.

[115] Nalewajski, R. F. Information principles in the theory of electronic structure. *Chem. Phys. Lett.*, 372:28, 2003.

[116] Á. Nagy. Fisher information in density functional theory. *J. Chem. Phys.*, 119:9401, 2003.

[117] Nagy, Á. Fisher information in a two-electron entangled artificial atom. *Chem. Phys. Lett.*, 425:154, 2006.

[118] Nagy, Á.; Sen, K. D. Atomic Fisher information versus atomic number. *Phys. Lett. A*, 360:291, 2006.

[119] Romera, E.; Dehesa, J. S. The Fisher-Shannon information plane, an electron correlation tool. *J. Chem. Phys.*, 120:8906, 2004.

[120] Romera, E. Stam's principle D-dimensional uncertainty-like relationships and some atomic properties. *Mol. Phys.*, 100:3325, 2002.

[121] Shubin, L. On the relationship between densities of Shannon entropy and Fisher information for atoms and molecules. *J. Chem. Phys.*, 126(19):191107, 2007.

[122] Nagy, Á. Fisher information and steric effect. *Chem. Phys. Lett.*, 449:212, 2007.

[123] Nagy, Á.; Liu, S. Local wave-vector, Shannon and Fisher information. *Phys. Lett. A*, 372:1654, 2008.

[124] Szabó, J. B.; Sen, K. D.; Nagy, Á. The FisherShannon information plane for atoms. *Phys. Lett. A*, 372:2428, 2008.

[125] Ou, J.-H.; Ho, Y. K. Shannon Information Entropy in Position Space for the Ground and Singly Excited States of Helium with Finite Confinements. *Atoms*, 5:15, 2017.

[126] Rodriguez-Bautista, M.; Vargas, R.; Aquino, N.; Garza, J. Electron-density delocalization in many-electron atoms confined by penetrable walls: A Hartree-Fock study. *Int. J. Quant. Chem.*, 118(13):e25571, 2018.

[127] López-Rosa, S.; Martín, A. L.; Antolín, J.; Angulo, J. C. Electron-pair entropic and complexity measures in atomic systems. *Int. J. Quant. Chem.*, 119:e25861, 2019.

[128] Hendeković, J.; Kučar J. Reply to the comment on the energy variation method. *Chem. Phys. Lett.*, 100:117, 1983.

[129] Pathak, R. K.; Bhattacharyya K. On the nonlinear manifold energy variation method and excited state calculations. *Chem. Phys. Lett.*, 230:437, 1994.

[130] Jelic, V.; Marsiglio F. The double-well potential in quantum mechanics: a simple, numerically exact formulation. *Int. J. Quant. Chem.*, 33:1651, 2012.

[131] Roy, A. K. Calculation of the spiked harmonic oscillators through a generalized pseudospectral method. *Phys. Lett. A*, 321:231, 2004.

[132] Roy, A. K.; Jalbout, A. F. Bound states of the generalized spiked harmonic oscillator. *J. Mol. Struct. (Theochem)*, 853:27, 2008.

[133] Roy, A. K. Calculation of the bound states of power-law and logarithmic potentials through a generalized pseudospectral method. *J. Phys. G*, 30:269, 2004.

[134] Roy, A. K. The generalized pseudospectral approach to the bound states of Húlthen and Yukawa potential. *Pramana-J. Phys.*, 65:1, 2005.

[135] Roy, A. K.; Jalbout, A. F.; Proynov, E. I. Bound state spectra of the 3D rational potential. *Int. J. Quant. Chem.*, 108:827, 2008.

[136] Roy, A. K.; Jalbout, A. F.; Proynov, E. I. Accurate calculation of the bound states of hellmann potential. *J. Math. Chem.*, 44:260, 2008.

[137] Roy, A. K. Studies on some exponentialscreened coulomb potentials. *Int. J. Quant. Chem.*, 113:1503, 2013.

[138] Roy, A. K. Accurate ro-vibrational spectroscopy of diatomic molecules in a Morse oscillator potential. *Results in Physics*, 3:103, 2013.

[139] Roy, A. K. Studies on the bound-state spectrum of Hyperbolic potential. *Few-Body Systems*, 55:143, 2014.

[140] Roy, A. K. Ro-vibrational studies of diatomic molecules in a shifted Deng-Fan oscillator potential. *Int. J. Quant. Chem.*, 114:383, 2014.

[141] Roy, A. K. Ro-vibrational spectroscopy of molecules represented by Tietz-Hua oscillator potential. *J. Math. Chem*, 52:1405, 2014.

[142] Roy, A. K. Studies on some singular potentials in quantum mechanics. *Int. J. Quant. Chem.*, 104:861, 2005.

[143] Roy, A. K.; Chu, S.-I. Density-functional calculations on singly and doubly excited Rydberg states of many-electron atoms. *Phys. Rev. A*, 65:052508, 2002.

[144] Roy, A. K.; Chu, S. I. Quantum fluid dynamics approach for strong field processes: Application to the study of multiphoton ionization and high-order harmonic generation of He and Ne in intense laser fields. *Phys. Rev. A*, 65:043402, 2002.

[145] Roy, A. K.; Chu, S. I. Quantum fluid dynamics approach for electronic structure calculation: Application to the study of ground-state properties of rare gas atoms. *J. Phys. B*, 35:2075, 2002.

[146] Roy, A. K. Studies on the hollow states of atomic lithium using a density functional approach. *J. Phys. B*, 37:4569, 2004.

[147] Roy, A. K. Density functional studies on the hollow resonances in Li-isoelectronic sequence (Z=4-10). *J. Phys. B*, 38:1591, 2005.

[148] Roy, A. K.; Jalbout, A. F. Ground and excited states of Li^-, Be^- through a density-based approach. *Chem. Phys. Lett.*, 445:355, 2007.

[149] Sahni, V.; Li, Y.; Harbola, M. K. Atomic structure in the Pauli-correlated approximation. *Phys. Rev. A*, 45:1434, 1992.

[150] Brual, G.; Rothstein, S. M. Rare gas interactions using an improved statistical method. *J. Chem. Phys.*, 69:1177, 1978.

[151] Lee, C.; Yang, W.; Parr, R. G. Development of the Colle-Salvetti correlation-energy formula into a functional of the electron density. *Phys. Rev. B*, 37:785, 1988.

[152] Cover, T. M.; Thomas, J. A. *Elements of Information Theory*. John Wiley & Sons Ltd., New York, 2006.

[153] Nielsen, M. A.; Chuang, I. L. *Quantum Computation and Quantum Information*. Cambridge University Press, Cambridge, 2010.

[154] Ramírez, J. C.; Soriano, C.; Esquivel, R. O.; Sagar, R. P.; Hô, M.; Smith Jr., V. H. Jaynes information entropy of small molecules: Numerical evidence of the Colins conjecture. *Phys. Rev. A*, 56:4477, 1997.

[155] Delle, L. S. Shannon entropy and many-electron correlations: Theoretical concepts, numerical results, and Collins conjecture. *Int. J. Quant. Chem.*, 115:1396, 2015.

[156] Ghiringhelli, L. M.; Hamilton, I. P.; Delle, L. S. Interacting electrons, spin statistics, and information theory. *J. Chem. Phys.*, 132:014106, 2010.

[157] Ghiringhelli, L. M.; Delle, L. S.; Mosna, R. A.; Hamilton, I. P. Information-theoretic approach to kinetic-energy functionals: the nearly uniform electron gas. *J. Math. Chem.*, 48:78, 2010.

[158] Nagy, Á. Fisher and Shannon information in orbital-free density functional theory. *Int. J. Quant. Chem.*, 115:1392, 2014.

[159] Alipour, M. Making a happy match between orbital-free density functional theory and information energy density. *Chem. Phys. Lett.*, 635:210, 2015.

[160] Delle, L. S. On the scaling properties of the correlation term of the electron kinetic functional and its relation to the Shannon measure. *Europhys. Lett.*, 86:40004, 2009.

[161] Mohajeri, A.; Alipour, M. Information energy as an electron correlation measure in atomic and molecular systems. *Int. J. Quantum Inf.*, 07:801, 2009.

[162] Grassi, A. A relationship between atomic correlation energy of neutral atoms and generalized entropy. *Int. J. Quant. Chem.*, 111:2390, 2011.

[163] Flores-Gallegos, N. Informational energy as a measure of electron correlation. *Chem. Phys. Lett.*, 666:62, 2016.

[164] Alipour, M.; Badooei, Z. Toward electron correlation and electronic properties from the perspective of information functional theory. *J. Phys. Chem. A*, 122:6424, 2018.

[165] Barghathi, H.; Herdman, C.; Maestro, A. D. Rényi generalization of the accessible entanglement entropy. *Phys. Rev. Lett.*, 121:150501, 2018.

[166] Rényi, A. On measures of entropy and information. *Proc. Fourth Berkeley Symp. on Math. Stat. and Prob.*, page 547, 1961.

[167] Rényi, A. *Probability Theory.* North-Holland Pub. Company, Amsterdam, 1970.

[168] Varga, I.; Pipek, J. Rényi entropies characterizing the shape and the extension of the phase space representation of quantum wave functions in disordered systems. *Phys. Rev. E*, 68:026202, 2003.

[169] Renner, R.; Gisin, N.; Kraus. B. Information-theoretic security proof for quantum-key-distribution protocols. *Phys. Rev. A*, 72:012332, 2005.

[170] Lévay, P.; Nagy, S.; Pipek, J. Elementary formula for entanglement entropies of fermionic systems. *Phys. Rev. A*, 72:022302, 2005.

[171] Verstraete, F.; Cirac, J. I. Matrix product states represent ground states faithfully. *Phys. Rev. B*, 73:094423, 2006.

[172] Bialas, A.; Czyz, W.; Zalewski, K. Moments of the Wigner function and Rényi entropies at freeze-out. *Phys. Rev. C*, 73:034912, 2006.

[173] Salcedo, L. L. Phase-space localization of antisymmetric functions. *J. Math. Phys.*, 50:012106, 2009.

[174] Nagy, Á.; Romera. E. Maximum Rényi entropy principle and the generalized Thomas-Fermi model. *Phys. Lett. A*, 373:844, 2009.

[175] McMinis, J.; Tubman, N. M. Rényi entropy of the interacting Fermi liquid. *Phys. Rev. B*, 87:081108(R), 2013.

[176] Liu, S.-B.; Rong, C.-Y.; Wu, Z.-M.; Lu, T. Rényi entropy, Tsallis entropy and Onicescu information energy in Density Functional Reactivity Theory. *Acta. Phys. Chim. Sin.*, 31:2057, 2015.

[177] Kim, I. Rényi-α entropies of quantum states in closed form: Gaussian states and a class of non-Gaussian states. *Phys. Rev. E*, 97:062141, 2018.

[178] B. R. Frieden. *Science from Fisher Information: A Unification.* Cambridge University Press, Cambridge, 2004.

[179] Sen, K. D., editor. *Statistical Complexity: Applications in Electronic Structure*. Springer International Publishing, Netherland, 2011.

[180] Noughabi, H. A.; Chahkandi, M. Informational energy and its application in testing normality. *Ann. Data Sci.*, 2:391, 2015.

[181] Alcoba, D. R.; Torre, A.; Lain, L.; Massaccesi, G. E.; Oña, O. B.; Ayers, P. W.; Van Raemdonck, M.; Bultinck, P.; Van Neck, D. Performance of Shannon-entropy compacted N-electron wave functions for configuration interaction methods. *Theor. Chem. Acc.*, 135:153, 2016.

[182] Zozor, S.; Portesi, M.; Vignat, C. Some extensions of the uncertainty principle. *Physica A*, 387:4800, 2008.

[183] Landsberg, P. T.; Shiner; J. S. Disorder and complexity in an ideal non-equilibrium Fermi gas. *Phys. Lett. A*, 245:228, 1998.

[184] Shiner, J. S.; Landsberg, P. T.; Davison, M. Simple measure for complexity. *Phys. Rev. E*, 59:1459, 1999.

[185] Sen, K. D.; Antolín, J.; Angulo, J. C. Fisher-Shannon analysis of ionization processes and isoelectronic series. *Phys. Rev. A*, 76:032502, 2007.

[186] Sen, K. D.; Antolín, J.; Angulo, J. C. Fisher-Shannon plane and statistical complexity of atoms. *Phys. Rev. A*, 372:670, 2008.

[187] Calbet, X.; López-Ruiz, R. Tendency towards maximum complexity in a nonequilibrium isolated system. *Phys. Rev. E*, 63:066116, 2001.

[188] Martin, M. T.; Plastino, A.; Rosso, A. O. Generalized statistical complexity measures: Geometrical and analytical properties. *Physica A*, 369:439, 2006.

[189] Catalán, R. G.; Garay, J.; López-Ruiz, R. Features of the extension of a statistical measure of complexity to continuous systems. *Phys. Rev. E*, 66:11102, 2002.

[190] Sánchez, J. R.; López-Ruiz, R. A method to discern complexity in two-dimensional patterns generated by coupled map lattices. *Physica A*, 355:633, 2005.

[191] Angulo, J. C.; Antolín, J. Atomic complexity measures in position and momentum spaces. *J. Chem. Phys.*, 128:164109, 2008.

[192] Esquivel, R. O.; Angulo, J. C.; Antolín, J.; Dehesa, J. S.; López-Rosa, S.; Flores-Gallegos, N. Analysis of complexity measures and information planes of selected molecules in position and momentum spaces. *Phys. Chem. Chem. Phys.*, 12:7108, 2010.

[193] Welearegay, M. A.; Balawender, R.; Holas A. Information and complexity measures in molecular reactivity studies. *Phys. Chem. Chem. Phys.*, 16:14928, 2014.

[194] Esquivel, R. O.; Molina-Espíritu, M.; Angulo, J. C.; Antolín, J.; Flores-Gallegos, N.; Dehesa, J. S. Information-theoretical complexity for the hydrogenic abstraction reaction. *Mol. Phys.*, 109:2353, 2011.

[195] Molina-Espíritu, M.; Esquivel, R. O.; Angulo, J. C.; Antolín, J.; Dehesa, J. S. Information-theoretical complexity for the hydrogenic identity S_N2 exchange reaction. *J. Math. Chem.*, 50:1882, 2012.

[196] Esquivel, R. O.; Molina-Espíritu, M.; Dehesa, J. S.; Angulo, J. C.; Antolín, J. Concurrent phenomena at the transition region of selected elementary chemical reactions: An information-theoretical complexity analysis. *Int. J. Quant. Chem.*, 112:3578, 2012.

[197] Kullback, S.; Leibler, R. A. On information and sufficiency. *Ann. Math. Stat.*, 22:79, 1951.

[198] Kullback, S. *Information Theory and Statistics*. John Wiley & Sons, New York, 1959.

[199] Sagar, R. P.; Guevara, N. L. Relative entropy and atomic structure. *J. Mol. Struct. (Theochem)*, 857:72, 2008.

[200] Villani, C. *Topics in Optimal Transportation, in Graduate Studies in Mathematics*, volume 58. American Mathematical Society, New York, 2000.

[201] Yamano, T. Relative Fisher information for Morse potential and isotropic quantum oscillators. *J. Math. Phys.*, 54:113301, 2013.

[202] Sánchez-Moreno, P.; Zarzo, A.; Dehesa, J. S. Jensen divergence based on Fisher's information. *J. Phys. A*, 45:125305, 2012.

[203] Toscani, G. Poincaré-type inequalities for stable densities. *Ric. Mat.*, 66:15, 2017.

[204] Yamano, T. de bruijn-type identity for systems with flux. *Eur. Phys. J. B*, 86:363, 2013.

[205] Frieden, B. R.; Plastino, A.; Plastino, A. R.; Soffer, B. H. Fisher-based thermodynamics: Its Legendre transform and concavity properties. *Phys. Rev. E*, 60:48, 1999.

[206] Antolín, J.; Angulo, J. C.; López-Rosa, S. Fisher and Jensen-Shannon divergences: Quantitative comparisons among distributions. Application to position and momentum atomic densities. *J. Chem. Phys.*, 130:074110, 2009.

[207] Levämäki, H.; Nagy, Á; Vilja, I.; Kokko, K.; Vitos, L. Kullback-Leibler and relative Fisher information as descriptors of locality. *Int. J. Quant. Chem.*, 118:e25557, 2017.

[208] Flego, S. P.; Plastino, A.; Plastino, A. R. Non-linear relativistic diffusions. *Physica A*, 390:2776, 2011.

[209] Frieden, B. R.; Soffer, B. H. Weighted Fisher informations, their derivation and use. *Phys. Lett. A*, 374:3895, 2010.

[210] Flego, S. P.; Plastino, A.; Plastino, A. R. Fisher information, the Hellmann-Feynman theorem, and the Jaynes reciprocity relations. *Ann. Phys.*, 326:2533, 2011.

[211] Venkatesan, R. C.; Plastino, A. Legendre transform structure and extremal properties of the relative Fisher information. *Phys. Lett. A*, 378:1341, 2014.

[212] Venkatesan, R. C.; Plastino, A. Hellmann-Feynman connection for the relative Fisher information. *Ann. Phys.*, 359:300, 2015.

[213] Yamano, T. Relative Fisher information of hydrogen-like atoms. *Chem. Phys. Lett.*, 691:196, 2018.

[214] Sañudo, J.; López-Ruiz, R. Statistical complexity and Fisher-Shannon information in the H-atom. *Phys. Lett. A*, 372:5283, 2008.

[215] Mukherjee, N.; Roy, A. K. Quantum mechanical virial-like theorem for confined quantum systems. *Phys. Rev. A*, 99:022123, 2019.

[216] Roy, A. K.; Singh, R.; Deb, B. M. Density functional calculations on triply excited states of lithium isoelectronic sequence. *Int. J. Quant. Chem.*, 65(4):317, 1997.

[217] Roy, A. K.; Singh, R.; Deb, B. M. Density functional calculations for doubly excited states of He, Li^+, Be^{2+} and B^{3+} $(^{1,3}S^e, ^{1,3}P^o, ^{1,3}D^e, ^{1,3}F^o, ^1G^e)$. *J. Phys. B*, 30:4763, 1997.

[218] Roy, A. K.; Deb, B. M. Atomic inner-shell transitions: A density functional approach. *Phys. Lett. A*, 234:465, 1997.

[219] Roy, A. K.; Deb, B. M. Density functional calculations on neon satellites. *Chem. Phys. Lett.*, 292:461, 1998.

[220] Singh, R.; Roy, A. K.; Deb, B. M. Density functional calculations on low-lying singly excited states of open-shell atoms. *Chem. Phys. Lett.*, 296:530, 1999.

[221] Montgomery Jr., H. E.; Pupyshev, V. I. Confined helium: Excited singlet and triplet states. *Phys. Lett. A*, 377:2880, 2013.

[222] Garza, J.; Hernández-Pérez, J. M.; Ramírez, J.; Vargas, R. Basis set effects on the Hartree-Fock description of confined many-electron atoms. *J. Phys. B*, 45:015002, 2011.

Chapter 2

Information Entropy of Molecular Ensembles: Recent Advances and Prospectives

Denis Sh. Sabirov*
Institute of Petrochemistry and Catalysis, UFRC RAS, Russia

Abstract

This mini-review demonstrates the application of the information-entropy formalism to the description of molecular ensembles. We have discussed most interesting results in this field obtained in our group: superadditivity of the information entropy of molecular ensembles, features of mixing molecular ensembles, and magic ensembles characterized with zero information entropy of mixing. We focus on cooperative information entropy, which is an emergent property resulted from mixing. The results considered in the review relate to this quantity. Also, the similarity of mathematical regularities underlying both information-entropy description of discrete chemical objects and quantum information processing has been noted. We assume novel modes of interacting these fields and provide some prospective of information-entropy-based discrete mathematical chemistry and related interdisciplinary studies.

Keywords: information entropy, cooperative information entopy, information entropy of mixing, molecular ensemble, molecular meta-ensemble, discrete information-entropy formalism, continual information-entropy formalism

---

* Corresponding Author's Email: diozno@mail.ru.

In: Understanding Information Entropy
Editor: Vijay Kumar
ISBN: 979-8-88697-839-1

1. Introduction

Information entropy is nowadays a concept widely adopted by natural sciences. Originally, it was introduced in classical works of Shannon and Hartley to the theory of communication (Shannon, 1948; Hartley 1928) but then found useful for solving scientific problems beyond this field. For example, physics was used information entropy for describing dynamical systems and non-equilibrium processes (Haken, 1989). These first applications revealed that the concept is efficient for numerical characterizing the objects, which have probabilistic nature or representable as sets (Sabirov and Shepelevich, 2021). Usually, chemistry is a discipline that follows physics in the adoption of mathematical concepts but not in this case. The discovery of nucleic acids and their structure associated with genetic information processing made the information entropy concepts facilely adoptable by biology. Digitalization of an object is often associated with its grammatization, viz. representing as words (Sabirov, 2021). The structure of nucleic acids can be represented as an alphabet, so it is closer to treating in terms of information entropy than the structures of typical (small) molecules (Sabirov and Shepelevich, 2021; Akhter et al., 2013). We think that biology has overtaken chemistry due to the features of representing chemical objects that arise from the multilayerness of chemical structure. A chemical object can be treated at molecular, atomic, or electron levels (Kerber et al., 2004; Musil et al., 2021). Hence, several modes of applying information entropy concepts arise. To avoid their spacious discussion, refer to the review (Sabirov and Shepelevich, 2021)), we simplify this issue as a direct correspondence between the levels of chemical structure and chemical applications (Table 1). Note the hierarchy of the chemical levels: to describe the molecules, one needs to describe its atoms and/or bonds; to describe the atoms, one needs to describe its electrons etc.

We believe that 'pure' chemical problems deal mainly with chemical structure and the classical chemical level of detailing the structure corresponds to its representing as sets of atoms and chemical bonds. Molecules are facilely represented as molecular graphs, to which information-theoretic analysis is applicable. In molecular graphs, vertices and edges correspond respectively to atoms and chemical bonds of molecules. Here occur two opportunities of the applications: considering the sets of atoms or the set of bonds. In this work, we prefer focus on atoms because the concept of chemical bond is fuzzy (Minkin, 2012) and its definition is vague and conventional rather than strictly mathematized (in contrast, atoms strictly correspond to the vertices of molecular graph if we deal with molecular topology and their spatial positions are determinable with experimental and computational techniques). This work corresponds to the second level of Table 1 and we have succeeded in the applications of the information-theoretical approach to interstellar molecules (Bertz, 2003), carbon (Sabirov, 2016; Sabirov et al., 2016, 2021) and oxygen molecular forms (Sabirov and Shepelevich, 2015), dendrimers (Sabirov and Koledina, 2020) and isentropic molecules (Sabirov et al., 2021). When applied, it allows discriminating most and least probable structures in the series of related compounds and this complementarily assists rationalizing available experimental data. Note that other groups applies similar approaches to natural compounds (Castellano et al., 2014; Castellano and Torrens, 2015), inorganic and molecular crystals (Krivovichev et al., 2018, 2022; Banaru et al., 2021; Banaru and Aksenov, 2022).

Table 1. The levels of chemical structure and features of the corresponding information-entropy applications

Level of detailing chemical structure	**Field of application**	**Major type of the used information entropy concepts**	**Type of analysis**	**Exemplifying references**
Molecules	Signal processing	Discrete	A posteriori	(García-Garibay, 2010; Mac Fhionnlaoich and Guldin, 2020; Zhdanov, 1979; Kobozev, 1971)
Atoms and chemical bonds	Molecular complexity and QSAR studies	Discrete	A priori	(Zhdanov, 1979; Bonchev, 1995; Barigye et al., 2014; Basak et al., 1984, 1997; Bertz, 2003; Sabirov, 2016; Sabirov and Osawa, 2015; Ghorbani et al., 2018, 2021; Böttcher, 2016; Stankevich et al., 1988; Dehmer and Mowshowitz, 2011)
Electrons (electronic density)	Quantum chemistry and quantum mechanics	Continual	A priori	(Nalewajski, 2019, 2021; Nalewajski and Parr, 2000)

Information entropy descriptors used in structural chemistry deal mainly with their applications to individual molecules. At the same time, some chemical problems require describing the collectives of the molecules and the most important issue of this type is digitalizing chemical reactions, whereby one must describe jointly products and reactants. Of course, there were approaches that imply simple summation of the molecular values to obtain the collective one (see the first work of Karreman, 1955). However, such a simple summation leads to counterintuitive results not rationalized in chemical terms (Sabirov, 2018). Therefore, other methodologies are developed to correct the mismatch.

In this Chapter, we consider in brief our previous works on information entropy of molecular ensembles. Previously, we have demonstrated the efficiency of this new approach to solve chemical problems that require numerical describing the collectives of the molecules (Sabirov, 2018, 2020; Sabirov et al., 2022). This work focuses on some mathematical features of the parameters used in our approach.

2. Basic Definitions

The information entropy of each individual molecule (h_i) is calculated taking into account the number of types of nonequivalent atoms in its structure (n) and their populations (ΣN_j is the total number of atoms in the molecule):

$$h = -\sum_{j=1}^{n} \frac{N_j}{\sum_{j=1}^{n} N_j} \log_2 \frac{N_j}{\sum_{j=1}^{n} N_j} \tag{1}$$

The fraction of atoms belonging to each atom type of the molecule ($\Sigma\, \omega_j = 1$) equals:

$$\omega_j = \frac{N_j}{\sum_{j=1}^{n} N_j} \tag{2}$$

Further, we use the concept of molecular ensemble (ME) introduced in computational chemistry by (Ugi and Gillespie, 1971). Its definition is simple: ME is a set of the molecules treated computationally together. Herewith, we distinguish the concepts of molecular complex and molecular ensemble. Molecular complex is a new chemical entity composed from original chemical individuals. Molecular ensemble is just a set of chemical individuals (molecules), a mechanical mixture that does not form a new chemical entity. In mathematical aspect, molecular complex is usually described with one molecular graph whereas the set of molecular graphs corresponding to the constituting molecules is attributed to the molecular ensemble.

To have a chemical meaning, the information entropy of an ensemble h_{ME} of m molecules is calculated similarly to formula (1):

$$h_{\mathrm{ME}} = -\sum_{i=1}^{m}\sum_{j=1}^{n}\frac{N_{ij}}{N}\log_2\frac{N_{ij}}{N} \quad (3)$$

This expression does not explicitly connect the informational entropy of the molecular ensemble and its members. Therefore, (Sabirov, 2020) has transformed it to establish such connection:

$$h_{\mathrm{ME}} = H_{\Omega} + \sum_{i=1}^{m}\omega_i h_i \quad (4)$$

where, the first term is called the cooperative entropy emerging when molecules are combined into an ensemble:

$$H_{\Omega} = -\sum_{i=1}^{m}\omega_i \log_2 \omega_i \quad (5)$$

This is an emergent property of the ensemble as it is a parameter of the system (ME) but not its parts (molecules). We have shown that h_i values take into account the structural features of the molecules whereas ω_i do not and depend only on the size of the molecules. Consequently, H_{Ω} is defined by the molecular size (the numbers of atoms in the molecules). The view of Equation (5) is similar to typical information entropy of the set of m events in information theory (Nielsen and Chuang, 2009; Adzhemov and Sannikov, 2020), so function H_{Ω} inherits its properties. For example, its maximal value $H_{\Omega}^{max} = 1$ is achieved when all $\omega_i = 1/m$ (Nielsen and Chuang, 2009).

In chemical sense, it is important that cooperative information entropy reconciles the contradictions appeared under simple summation of information entropies of individual species (Sabirov, 2018, 2020). Furthermore, this chemical case demonstrates the superadditive behavior of the information entropy of complex systems when expressed through the information entropies of their parts.

(Sabirov, 2018, 2020) introduced another value associated with forming molecular ensembles: the information entropy of mixing is a measure of superadditivity of the information entropy of ME relative to the molecules:

$$\Delta h_{mix} = h_{\mathrm{ME}} - \sum_{i=1}^{m} h_i \tag{6}$$

It could be also expressed through the cooperative information entropy:

$$\Delta h_{mix} = H_{\Omega} - \sum_{i=1}^{m} (1 - \omega_i) h_i \tag{7}$$

3. Some Properties of Information Entropy of Molecular Ensemble

3.1. Information Entropy of Open Molecular Ensemble

We consider a dimorphic molecular ensemble, i.e., the ensemble consisting of the molecules of two types, A and B. Its openness means that it can grow by accepting additional molecules from the environment (Tukhbatullina et al., 2022). Let us put for simplicity that the ensemble initially contains n_{A} molecules A, which have size N_{A}, and starts absorbing molecules B. Thus, we obtain a series of dimorphic ensembles $n_{\mathrm{A}}\mathrm{A} + n_{\mathrm{B}}\mathrm{B}$. Their h_{ME} and H_{Ω} obtained from expressions (4) and (5) are the functions of n_{B}:

$$h_{\mathrm{ME}}(n_{\mathrm{B}}) = H_{\Omega}(n_{\mathrm{B}}) + \frac{n_{\mathrm{A}}N_{\mathrm{A}}}{n_{\mathrm{A}}N_{\mathrm{A}}+n_{\mathrm{B}}N_{\mathrm{B}}} h_{\mathrm{A}} + \frac{n_{\mathrm{B}}N_{\mathrm{B}}}{n_{\mathrm{A}}N_{\mathrm{A}}+n_{\mathrm{B}}N_{\mathrm{B}}} h_{\mathrm{B}} \tag{8}$$

$$H_{\Omega}(n_B) = -\frac{n_{\mathrm{A}}N_{\mathrm{A}}}{n_{\mathrm{A}}N_{\mathrm{A}}+n_{\mathrm{B}}N_{\mathrm{B}}} \log_2 \frac{n_{\mathrm{A}}N_{\mathrm{A}}}{n_{\mathrm{A}}N_{\mathrm{A}}+n_{\mathrm{B}}N_{\mathrm{B}}} - \frac{n_{\mathrm{B}}N_{\mathrm{B}}}{n_{\mathrm{A}}N_{\mathrm{A}}+n_{\mathrm{B}}N_{\mathrm{B}}} \log_2 \frac{n_{\mathrm{B}}N_{\mathrm{B}}}{n_{\mathrm{A}}N_{\mathrm{A}}+n_{\mathrm{B}}N_{\mathrm{B}}} \tag{9}$$

It is interesting to scrutinize extremums of the functions. Function $H_{\Omega}(n_B)$ achieves its maximal value, equal to 1, when $\omega_{\mathrm{A}} = \omega_{\mathrm{B}} = ½$. The latter corresponds to the number of molecules B in the ensemble equal to:

$$n_{\mathrm{B}}^{max} = \frac{n_{\mathrm{A}}N_{\mathrm{A}}}{N_{\mathrm{B}}} \tag{10}$$

In the case of function $h_{\mathrm{ME}}(n_{\mathrm{B}})$, the maximum is located at:

$$h_{\mathrm{ME}}^{max} = h_{\mathrm{A}} + \log_2(2^{h_{\mathrm{B}}-h_{\mathrm{A}}} + 1) \quad (11)$$

$$n_{\mathrm{B}}^{max} = \frac{n_{\mathrm{A}}N_{\mathrm{A}}}{N_{\mathrm{B}}} \times 2^{h_{\mathrm{B}}-h_{\mathrm{A}}} \quad (12)$$

The number of molecules B at the maximum corresponds to their fraction in the molecular ensemble expressed as:

$$\omega_{\mathrm{B}}^{max} = \frac{2^{h_{\mathrm{B}}-h_{\mathrm{A}}}}{2^{h_{\mathrm{B}}-h_{\mathrm{A}}+1}} \quad (13)$$

Thus, the fraction of molecule B at h_{ME}^{max} depends on the information entropies of the ensemble's molecules whereas H_{Ω}^{max} always corresponds to $\omega_{\mathrm{B}}^{max}$ = ½. Maxima of both functions are sensitive to the ratio of the number of atoms in molecules A and B.

General views of typical functions $h_{\mathrm{ME}}(n_{\mathrm{B}})$ and $H_{\Omega}(n_B)$ are shown in Figure 1. We exemplify this section with three typical cases of $h_{\mathrm{A}} \neq h_{\mathrm{B}}$ (C_{70} + O_3), $h_{\mathrm{A}} = h_{\mathrm{B}}$ ($10C_2H_4$ + O_3), and $h_{\mathrm{A}} = 0$, $h_{\mathrm{B}} \neq 0$ (Table 2). Note that $h_{\mathrm{ME}} \rightarrow h_{\mathrm{B}}$ and $H_{\Omega} \rightarrow 0$ when $n_{\mathrm{B}} \rightarrow \infty$ in the open molecular ensemble. Generally, the maxima of these functions do not coincide. The coincidence takes place only in the case of isentropic molecules ($h_{\mathrm{A}} = h_{\mathrm{B}}$). In the case of zero-*h* molecules ($h_{\mathrm{A}} = h_{\mathrm{B}} = 0$) the plots $h_{\mathrm{ME}}(n_{\mathrm{B}}) = H_{\Omega}(n_B)$, i.e., the complexity of the ensemble is completely attributed to the emergent parameter.

Table 2. Locations of the maxima of two characteristic information-entropy functions describing the selected open molecular ensembles*

Molecular ensemble series**	**Function $H_{\Omega}(n_B)$**			**Function $h_{\mathrm{ME}}(n_{\mathrm{B}})$**		
	n_{B}^{max}	$\omega_{\mathrm{B}}^{max}$	H_{Ω}^{max}	n_{B}^{max}	$\omega_{\mathrm{B}}^{max}$	h_{ME}^{max}
$C_{70} + n_{\mathrm{B}}O_3$	23	0.5	1	9	0.28	2.723
$10O_3 + n_{\mathrm{B}}C_2H_4$	5	0.5	1	5	0.5	1.918
$C_{60} + n_{\mathrm{B}}O_3$	20	0.5	1	38	0.66	1.531

* See the plots in Figure 1.

** The information entropies of the isolated molecules: $h(C_{70}) = 2.236$, $h(O_3) = h(C_2H_4) = 0.918$, and $h(C_{60}) = 0$ bits.

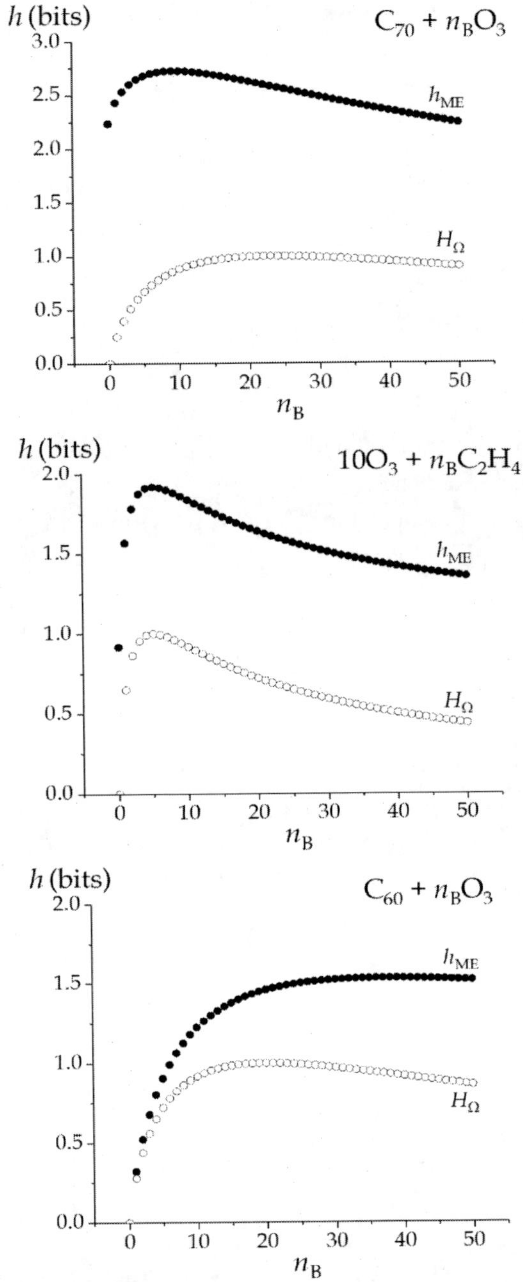

Figure 1. Information entropy functions of the selected open molecular ensembles (see also Table 2).

4. Mixing Molecular Ensembles

Mixing molecules is accompanied with emergence of the cooperative information entropy, which depends only on their size. Here, we do not discuss structural meaning of this emergent parameter just noting that its presence in Equation (4) is required for reliable description of the complexity of a molecular ensemble in terms of information-theoretic formalism (Sabirov, 2020). The cooperative information entropy should also emerge when more complex chemical species are mixed, e.g., when two ensembles form one molecular meta-ensemble. Such a case was previously described in brief. Here we consider a similar mixing to elucidate the relations between the emergent parameters of mixing on different hierarchic levels.

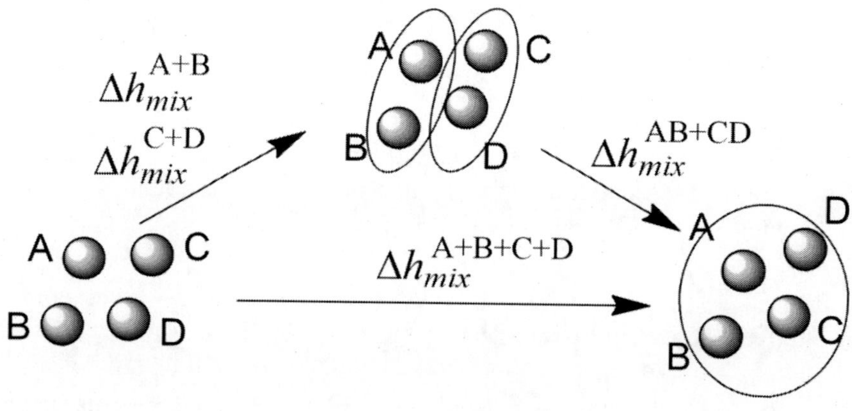

Figure 2. Scheme of formation meta-ensemble ABCD directly from the individual molecules and via two-stage mixing with intermediate forming ensembles AB and CD.

We consider the formation of tetramolecular meta-ensemble ABCD as an example (Figure 2). It can be formed directly from separate molecules A, B, C, and D or through the intermediate stage when molecules unit in couples AB and CD, which then form final ABCD. Hence, we have a two-stage mixing. Its information entropies of mixing of the two stages are linked with a simple relations:

$$\Delta h_{mix}^{\mathrm{A+B+C+D}} = \Delta h_{mix}^{\mathrm{A+B}} + \Delta h_{mix}^{\mathrm{C+D}} + \Delta h_{mix}^{\mathrm{AB+CD}} \quad (14)$$

The relations between the cooperative entropies of different mixing levels are connected more sophisticatedly. Before showing the relation, we introduce a novel fraction designation:

$$\omega_{\mathrm{AB}}^{\mathrm{ABCD}} = \frac{N_{\mathrm{A}}+N_{\mathrm{B}}}{N_{\mathrm{A}}+N_{\mathrm{B}}+N_{\mathrm{C}}+N_{\mathrm{D}}} \qquad (15)$$

that means the fraction of ensemble AB in meta-ensemble ABCD (and similar for CD in ABCD). Further, we apply Equation (5) to different steps of our mixing scheme and make some simplifications taking into account that:

$$\omega_{\mathrm{A}}^{\mathrm{ABCD}} = \omega_{\mathrm{A}}^{\mathrm{AB}} \omega_{\mathrm{AB}}^{\mathrm{ABCD}} \qquad (16)$$

and similar for B, C, and D. Then we obtain:

$$H_{\Omega}^{\mathrm{ABCD}} = H_{\Omega}^{\mathrm{AB+CD}} + \omega_{\mathrm{AB}}^{\mathrm{ABCD}} H_{\Omega}^{\mathrm{A+B}} + \omega_{\mathrm{CD}}^{\mathrm{ABCD}} H_{\Omega}^{\mathrm{C+D}} \qquad (17)$$

Thus, in contrast to the information entropy of mixing, the cooperative entropy of meta-ensemble additionally depends on the ME fractions attributed to the intermediate hierarchic level.

5. Magic Molecular Ensembles

We came to special case of molecular ensembles when studying the factors defining the sign of the information entropy of mixing molecules. According to Equation (7), H_Ω increases Δh_{mix} whereas $\Sigma(1 - \omega_i)h_i$ decreases. The last sum becomes zero if $h_i = 0$, which may be both when mixed molecules are the same ($H_\Omega = 0$ and $\Delta h_{mix} = 0$) or not ($\Delta h_{mix} = H_\Omega$). Otherwise, the sign of Δh_{mix} is a result of the balance between H_Ω and $\Sigma(1 - \omega_i)h_i$. Based on various examples, we have shown that Δh_{mix} is able to obtain negative and positive values (Sabirov, 2020).

One can mistakenly consider that $\Delta h_{mix} = 0$ if any identical molecules are mixed. However, this is a mental pitfall caused with the fallacious analogy with thermodynamic entropy.

Indeed, if we consider the ensemble of m identical molecules and apply Equations (4) and (6), we obtain that

$h_{\mathrm{ME}} = h$ and $\Delta h_{mix} = (1 - m)h$.

This result have information-theoretic interpretation. When m identical molecules are separated, we need h bits per each molecule to describe their structures or mh in total. If the molecules are united in one ensemble having $h_{ME} = h$, the similarity reduces the required information resources to code the structures (Sabirov, 2020). Here, $\Delta h_{mix} = 0$ only in the trivial case of mixing identical zero-h molecules.

Further, we consider a molecular ensemble of m isentropic molecules (with entropies equal to h) having the partition pattern proportional to the molecular composition of the ensemble. In other words, the distribution of the atoms over the molecules in the ensemble is proportional to the distribution of the atoms over the atom types in the molecules. Accordingly, the molecular ensemble consists of the molecules having N_1, $N_2 = k_2N_1$, $N_3 = k_3N_1$, …, $N_n = k_nN_1$ atoms (k_i are natural). The molecules are ordered in line with non-decreasing size. The internal structure of the first (smallest) molecule of the set implies the partition of atoms over atom types having $N_{11} = N_{11}$, $N_{12} = k_2N_{11}$, $N_{13} = k_3N_{11}$, …, $N_{1n} = k_nN_{11}$ atoms (similarly ordered with non-decreasing size); the second molecule's partition is $N_{21} = k_2N_{11}$, $N_{22} = k_2^2N_{11}$, $N_{23} = k_2k_3N_{11}$, …, $N_{2n} = k_2k_nN_{11}$. The following matrix corresponds to the notations above:

$$N_{11}\begin{pmatrix} 1 & k_2 & k_3 & \dots & k_n \\ k_2 & k_2^2 & k_2k_3 & \dots & k_2k_n \\ k_3 & k_2k_3 & k_3^2 & \ddots & k_3k_n \\ \vdots & \vdots & \vdots & \ddots & \vdots \\ k_n & k_2k_n & k_3k_n & \cdots & k_n^2 \end{pmatrix} \tag{18}$$

Summation of the elements in the n-th row gives the number of atoms in the corresponding molecule. Molecular ensembles, which structure fits into matrix (18), demonstrate remarkable relations: they have $H_\Omega = h$ and $\Sigma\omega_ih_i = h$. Hence, the information entropy of such magic ensemble is independent of the number of the constituting molecules and equal to:

$$h_{\text{ME}} = 2h \tag{19}$$

The corresponding information entropy of mixing molecules equals:

$$\Delta h_{mix} = (2 - n)h \tag{20}$$

If the ensemble is bimolecular ($n = 2$), the information entropy of mixing becomes zero. An example of magic ME is shown in Figure 3. Currently, we have no chemical interpretation of such magic. Concerning to the example, ensemble $C_2H_4 + O_3$ obeys conditions (18)–(20) but chemically very similar system $C_3H_6 + O_3$ has no magic properties (Table 3). Thus, we have no definite answer whether the mentioned conditions have chemical sense or they are just a mathematical artefact. Nevertheless, we think that this result may be interesting in the context of other systems (beyond molecular ensembles), which have complex hierarchical structure and the original structural pattern (like partitions of the molecules over atom types in our case) is reproduced on the next hierarchical level (partition of the molecular ensemble over molecules).

Table 3. Information entropy parameters of two related molecular ensembles

Molecular Ensemble	**Information entropies of the molecules (h_i)**	**Fractions of the molecules (ω_i)**	**H_Ω**	**Δh_{mix}**	**h_{ME}**
$C_2H_4 + O_3$ (magic)	0.918 + 0.918	0.67 + 0.33	0.918	0	1.836
$C_3H_6 + O_3$ (ordinary)	2.642 + 0.918	0.75 + 0.25	0.811	–0.538	3.022

6. Prospective

6.1. Toward Information-Entropy Approach of Chemical Reactions

The developed approach allows a new mode of describing chemical reactions. A chemical reaction can be considered as the isomerization of one molecular ensemble into another one (Ugi and Gillespie, 1971). Hence, the change in information entropy at chemical reaction is calculated as the difference between the h_{ME} values corresponding to the initial and final molecular ensembles:

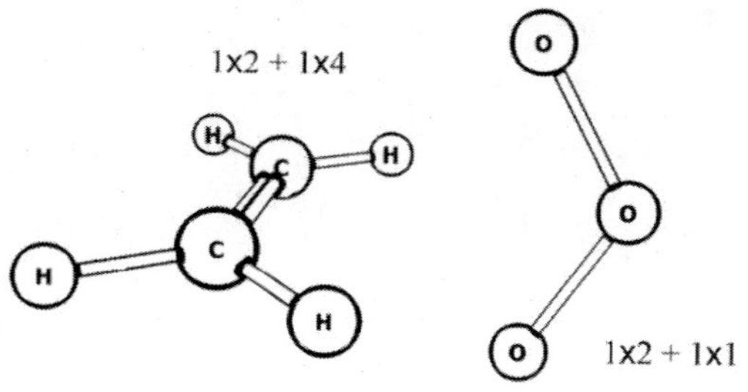

ethylene + ozone

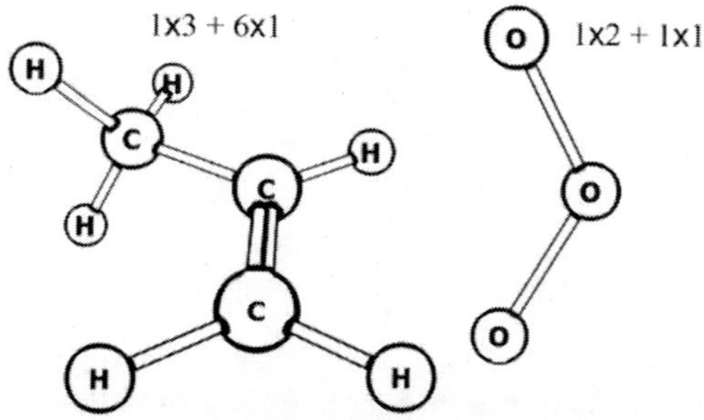

propylene + ozone

Figure 3. Two related molecular ensembles ethylene + ozone ($C_2H_4 + O_3$) and propylene + ozone ($C_3H_6 + O_3$). The partitions of the molecules are shown according to the conventional designtation [number of atom types]×[number of populating atoms]. The partition of only magic ensemble $C_2H_4 + O_3$ corresponds to condition (18).

$$\Delta h_{\mathrm{R}} = h_{\mathrm{ME}}^{prod} - h_{\mathrm{ME}}^{react} \tag{21}$$

After expressing h_{ME} values via Equation (4) and mathematical transformations, it is possible to separate the contributions to the information entropy of chemical process called redistribution and reorganization entropies:

$$\Delta h_{\mathrm{R}} = H_{redisrt} + H_{reorg} \tag{22}$$

The first term depends only on molecular size of the reacting and formed molecules (expressed with ω_i that define H_Ω):

$$H_{redistr} = H_{\Omega}^{prod} - H_{\Omega}^{react} \tag{23}$$

The second term of Equation (22) depends on both molecular size and molecular structure (ω_i and h_i values, respectively):

$$H_{reorg} = \sum_i^{prod} \omega_i h_i - \sum_j^{react} \omega_j h_j \tag{24}$$

In the reorganization entropy, we can select a part defined only by the structure of the reaction participants:

$$H_{reorg}^{str} = \sum_i^{prod} h_i - \sum_j^{react} h_j \tag{25}$$

The remaining part of H_{reorg} is unavoidably depends on ω_i and h_i:

$$H_{reorg}^{str+size} = \sum_j^{react}(1 - \omega_j)h_j - \sum_i^{prod}(1 - \omega_i)h_i \tag{26}$$

Thus, we obtain opportunity of the information-theoretic analysis of chemical reactions with discriminating changes associated with molecular size and molecular structure. The new approach was successfully tested for the digital description of classical reactions of organic compounds (reactions of atomization, isomerization, Diels–Alder, etc.) (Sabirov, 2020; Sabirov et al., 2022). It provides different estimates depending on the chemical nature of the interacting substances and resulting products, i.e., its use allows digital classification of chemical reactions. For example, the information entropy of etherification of alcohols depends on whether the same or different alcohol molecules react (Sabirov, 2020); the informational entropy of fullerene oligomerization is sensible to the parity of the number of monomeric units and addition pattern of the resulting nano-aggregates (Sabirov, 2017) etc.

6.2. Toward New Information-Theoretic Quantities for Analyzing Chemical Objects

Analyzing other works in the field of information processing, we found similarities and direct coincidence between the expressions used for describing molecular systems and quantum information processing. For example, Equation (6) is similar to the von Neumann entropy of mixed quantum states (see §§ 11.3.2 and 11.3.6 of book (Nielsen and Chuang, 2009)). For example, from the theorem of the joint entropy (Nielsen and Chuang, 2009):

$$S(\sum_i p_i |i\rangle\langle i| \oplus \rho_i) = H(p_i) + \sum_i p_i S(\rho_i) \tag{27}$$

and the analogy between the values of Equations (4) and (27) is obvious (p_i are arbitrary probability distribution, $|i\rangle$ is the set of the orthogonal states of system A, and ρ_i are the same for another system B).

In our approach, we apply information-entropy formalism to discrete chemical objects (molecules are represented as sets of the atoms). The discrete Shannon entropy has a continual analogue. Here, Equation (1) is rewritten as:

$$h = -\int_0^\infty p(r) \log p(r)\, dr \tag{28}$$

that corresponds to the continual probability distribution with random variable r:

$$\int_0^\infty p(r) dr = 1 \tag{29}$$

Similarly, we can rewrite expression (4):

$$h = \int_0^\infty \omega(r) h(r) dr - \int_0^\infty \omega(r) \log \omega(r)\, dr \tag{30}$$

As far as we can judge, Equation (30) has not been applied in mathematical chemistry. Taking into account that its discrete form found chemical applications, we propose that this Equation (and other quantities from the quantum information processing) may be useful for solving chemical problems.

6.3. Toward Interdisciplinary Studies on Interplays of Chemistry, Physics, and Information Theory

Our century is focused on interdisciplinary studies. Usually, disciplines interact spontaneously giving birth to new scientific directions on a disciplines' interface but we can predict the points of attitude, which have a high potential to induce interdisciplinary studies. We assume that information entropy and related quantities can be such a point.

Recursions and recursivity as self-reproducing patterns attracts researches from different fields. Philosophers provide an important role to recursivity in the structure of reality (Hui, 2019). Something similar we obtain in our case for objects of chemical reality when meeting magic molecular ensembles. Information-entropy formalism may provide here mathematical tools for describing the connectedness of patterns of different hierarchic levels of the organization of matter. Note that the attempts to find the mathematical regularities between the patterns has a long story (e.g., see work of Turing, 1952 on chemical morphogenesis).

In the modern philosophy of science, some extreme directions arise like Qbism, which describes all phenomena in terms of information processing (Yakovlev, 2020; Wheeler, 1990; Glick, 2021). In this direction, information has an independent existence, which is not reduced to energy or matter. If so, there is a problem of finding quantitative regularities between the information, matter, and energy. Works in this direction are performed (see, e.g., works on statistical (Kobozev, 1971; Parrondo et al., 2015; Sagawa, and Ueda, 2011) and geometrical thermodynamics (Parker and Jeynes, 2020), links between information and configurational entropies (Champion and Thurieau, 2020; Feng and Zhuang, 2020)).

Information theory is used in nontrivial interpretations of quantum mechanics (Zeilinger, 1999; Brukner and Zeilinger, 2009). As shown in the previous section, chemical structures and quantum phenomena are mathematically formalized in the same manner, so we expect closer interactions of these disciplines.

The same mathematical formalism for probability of emerging pattern, *h*-based structural descriptors, and assessing resources required for coding information can make information entropy a starting point for interdisciplinary research on the interfaces of physics, chemistry, and information sciences (Sabirov and Shepelevich, 2021). Such research can also involve the works whereby chemical objects are studied in terms of the theory of algebraic sets (Sokolov, 1991) or representing chemical reactions as sentences (the premises comes from incorporating linguophilosophic Wittgenstein's approach to rationalizing physical reality (Pris,' 2022) and works on representing chemical information (Hoffmann and Laszlo, 1991)).

At the end, we should note the disadvantages of the quantitative information theory as applied to chemical objects. The most significant disadvantage deals with ignoring the sense of the message focusing on its symbolic contents. Due to this, dissimilar chemical objects may obtain the same information-entropy estimates (Sabirov and Shepelevich, 2021; Sabirov and Koledina, 2020). In mathematical chemistry, this situation is corrected by invoking other (non-entropy) parameters or aprioral narrowing the semantic field of the study (e.g., considering only isomers, homologues, or related compounds). In this regard, the breakthrough is expected from the semantic theory of information (Kolchinsky and Wolpert, 2018) that could overcome the disadvantages of the quantitative one.

Conclusion

We have briefly reviewed interesting issues of the application of information entropy to molecular ensembles, discrete chemical objects composed with the molecules computationally treated together. Unusualness of such cases as superadditivity of the information entropy of molecular ensembles, formation of meta-ensembles, and magic ensembles characterized with zero information entropy of mixing arise due to the cooperative information entropy, a property emerging upon mixing molecules when forming ensemble.

We demonstrate how our findings fit into the modern tendencies of information science assuming ways of the future development of the information entropy applications in chemistry (digital description of chemical reactions and coming novel quantities from quantum information processing to mathematical chemistry) and their incorporating into the relevant interdisciplinary studies. Here we point the magic molecular ensembles, which case may be interesting beyond chemistry, as they demonstrate zero information of mixing and their structure represents a pattern self-reproducing on different levels of chemical structure.

Acknowledgments

This research was funded by the Russian Science Foundation, project "Information entropy of chemical reactions: A novel methodology for digital organic chemistry," grant number 22-13-20095.

References

Adzhemov, A. S. and Sannikov, V. G. (2020) *General Theory of Communication.* Goryachaya Liniya – Telekom, Moscow.

Akhter, S., Bailey, B. A., Salamon, P., Aziz, R. and Edwards, R. A. (2013) Applying Shannon's information theory to bacterial and phage genomes and metagenomes. *Scientific Reports*, 3, 1033.

Banaru, A. M. and Aksenov, S. M. (2022) Complexity of molecular nets: Topological approach and descriptive statistics. *Symmetry*, 14, 220.

Banaru, A., Aksenov, S. and Krivovichev, S. (2021) Complexity parameters for molecular solids. *Symmetry*, 13, 1399.

Barigye, S. J., Marrero-Ponce, Y., Pérez-Giménez, F. and Bonchev, D. (2014) Trends in information theory-based chemical structure codification. *Molecular Diversity*, 18, 673-686.

Basak, S. C., Gute, B. D. and Grunwald, G. D. (1997) Use of topostructural, topochemical, and geometric parameters in the prediction of vapor pressure: A hierarchical QSAR approach. *Journal of Chemical Information and Computer Sciences*, 37, 651-655.

Basak, S., Harriss, D. and Magnuson, V. (1984) Comparative study of lipophilicity versus topological molecular descriptors in biological correlations. *Journal of Pharmaceutical Sciences*, 73, 429-437.

Bertz, S. H. (2003) Complexity of synthetic reactions. The use of complexity indices to evaluate reactions, transforms and disconnections. *New Journal of Chemistry*, 27, 860-869.

Bonchev, D. (1995) Kolmogorov's information, Shannon's entropy, and topological complexity of molecules. *Bulgarian Chemical Communications*, 28, 567-582.

Böttcher, T. (2016) An additive definition of molecular complexity. *Journal of Chemical Information and Modeling*, 56, 462-470.

Castellano, G. and Torrens, F. (2015) Information entropy-based classification of triterpenoids and steroids from Ganoderma. *Phytochemistry*, 116, 305-313.

Castellano, G., Lara, A. and Torrens, F. (2014) Classification of stilbenoid compounds by entropy of artificial intelligence. *Phytochemistry*, 97, 62-69.

Dehmer, M. and Mowshowitz, A. (2011) A history of graph entropy measures. *Information Sciences*, 181, 57-78.

García-Garibay, M. A. (2010) The entropic enlightenment of organic photochemistry: Strategic modifications of intrinsic decay pathways using an information-based approach. *Photochemical & Photobiological Sciences*, 9, 1574-1588.

Ghorbani, M., Dehmer, M. and Zangi, S. (2018) Graph operations based on using distance-based graph entropies. *Applied Mathematics and Computation*, 333, 547-555.

Ghorbani, M., Rajabi-Parsa, M., Majidi, R. and Mirzaie, R. A. (2021) Novel results on entropy-based measures of fullerenes. *Fullerenes, Nanotubes and Carbon Nanostructures*, 29, 114-125.

Glick, D. (2021) QBism and the limits of scientific realism. *European Journal for Philosophy of Science*, 11, 53.

Haken, H. (1989) *Information and Self-Organization: A Macroscopic Approach to Complex Systems*. Springer, New York.

Hartley, R. V. L. (1928) Transmission of information. *The Bell System Technical Journal*, 7, 535-563.

Hui, Y. (2019) *Recursivity and Contingency*. Rowman & Littlefield International, London, New York.

Karreman, G. (1955) Topological information content and chemical reactions. *The Bulletin of Mathematical Biophysics*, 17, 279-285.

Kerber, A., Laue, R., Meringer, M. and Rücker, C. (2004) Molecules in silico: The generation of structural formulae and its applications. *Journal of Computer Chemistry, Japan*, 3, 85-96.

Kobozev, N. I. (1971) *A Study on Thermodynamics of the Information and Thinking Processes*. Moscow University Press, Moscow.

Krivovichev, S. V., Krivovichev, V. G. and Hazen, R. M. (2018) Structural and chemical complexity of minerals: Correlations and time evolution. *European Journal of Mineralogy*, 30, 231-236.

Krivovichev, S. V., Krivovichev, V. G., Hazen, R. M., Aksenov, S. M., Avdontceva, M. S., Banaru, A. M., Gorelova, L. A., Ismagilova, R. M., Kornyakov, I. V., Kuporev, I. V., Morrison, S. M., Panikorovskii, T. L. and Starova, G. L. (2022) Structural and chemical complexity of minerals: an update. *Mineralogical Magazine*, 86, 183-204.

Mac Fhionnlaoich, N. and Guldin, S. (2020) Information entropy as a reliable measure of nanoparticle dispersity. *Chemistry of Materials*, 32, 3701-3706.

Minkin, V. I. (2012) Current trends in the development of A. M. Butlerov's theory of chemical structure. *Russian Chemical Bulletin*, 61, 1265-1290.

Musil, F., Grisafi, A., Bartók, A. P., Ortner, C., Csányi, G. and Ceriotti, M. (2021) Physics-inspired structural representations for molecules and materials. *Chemical Reviews*, 121, 9759-9815.

Nalewajski, R. (2021) Resultant Information descriptors, equilibrium states and ensemble entropy. *Entropy*, 23, 483.

Nalewajski, R. F. (2019) Understanding electronic structure and chemical reactivity: Quantum-information perspective. *Applied Science*, 9, 1262.

Nalewajski, R. F. and Parr R. G. (2000) Information theory, atoms in molecules, and molecular similarity. *Proceedings of the National Academy of Sciences of the United States of America*, 97, 8879-8882.

Nielsen, M. A. and Chuang, I. L. (2009) *Quantum Computation and Quantum Information*. Cambridge University Press, Cambridge.

Pris', I. E. (2022) *Knowledge in a Context*; Aleteya: Saint Petersburg, Russia.

Sabirov, D. and Koledina, K. (2020) Classification of isentropic molecules in terms of Shannon entropy. *EPJ Web of Conferences*, 244, 01016.

Sabirov, D. and Shepelevich, I. (2015) Information entropy of oxygen allotropes. A still open discussion about the closed form of ozone. *Computational and Theoretical Chemistry*, 1073, 61-66.

Sabirov, D. S. (2016) Information entropy of interstellar and circumstellar carbon-containing molecules: Molecular size against structural complexity. *Computational and Theoretical Chemistry*, 1097, 83-91.

Sabirov, D. S. (2017) Information entropy change in [2+2]-oligomerization of the C_{60} fullerene. *International Journal of Chemical Modeling*, 9, 203-213.

Sabirov, D. S. (2018) Information entropy changes in chemical reactions. *Computational and Theoretical Chemistry*, 1123, 169-179.

Sabirov, D. S. (2020) Information entropy of mixing molecules and its application to molecular ensembles and chemical reactions. *Computational and Theoretical Chemistry*, 1187, 112933.

Sabirov, D. S. (2021) Digital chemistry: a novel branch of modern mathematical chemistry. *Vestnik Bashkirskogo Universiteta*, 26, 27-32.

Sabirov, D. S. and Osawa, E. (2015) Information entropy of fullerenes. *Journal of Chemical Information and Modeling*, 55, 1576-1584.

Sabirov, D. S. and Shepelevich, I. S. (2021) Information entropy in chemistry: An overview. *Entropy*, 23, 1240.

Sabirov, D. S., Ori, O., Tukhbatullina, A. A. and Shepelevich, I. S. (2021) Covalently bonded fullerene nano-aggregates $(C_{60})_n$: Digitalizing their energy–topology–symmetry. *Symmetry*, 13, 1899.

Sabirov, D. S., Sokolov, V. I. and Terentyev, O. A. (2016) Activation energies and information entropies of helium penetration through the fullerene walls. Insights into the formation of endofullerenes nX@$C_{60/70}$ ($n = 1$ and 2) from the information entropy approach. *RSC Advances*, 6, 72230-72237.

Sabirov, D. S., Tukhbatullina, A. A. and Shepelevich, I. S. (2022) Molecular size and molecular structure: Discriminating their changes upon chemical reactions in terms of information entropy. *Journal of Molecular Graphics and Modelling*, 110, 108052.

Sabirov, D., Tukhbatullina, A. and Shepelevich, I. (2021) Information entropy of regular dendrimer aggregates and irregular intermediate structures. *Liquids*, 1, 25-35.

Shannon, C. E. (1948) A mathematical theory of communication. *The Bell System Technical Journal*, 27, 379-423.

Stankevich, I. M., Stankevich, I. V. and Zefirov, N. S. (1988) Topological indices in organic chemistry. *Russian Chemical Reviews*, 57, 191-208.

Tukhbatullina, A. A., Shayakhmetova, R. R. and Sabirov, D. S. (2022) Cooperative information entropy of a dimorphic molecular ensemble of open type. *Vestnik Bashkirskogo Universiteta*, 27, 352-354.

Turing, A. (1952) The chemical basis of morphogenesis. *Philosophical Transactions of the Royal Society of London. Series B, Biological Sciences*, 237, 37-72.

Ugi, I. and Gillespie, P. (1971) Representation of chemical systems and interconversions bybe matrices and their transformation properties. *Angewandte Chemie, International Edition*, 10, 914-915.

Wheeler, J. A. (1990) Information, physics, quantum: The search for link. In: *Complexity, Entropy and the Physics of Information*. Ed.: W.H. Zurich. Addison-Wesley, Redwood City.

Yakovlev, V. A. (2020) Metaphysics in philosophy and science. In: *Metaphysics and Postmetaphysical Thinking*. Akademichesky Proekt, Moscow.

Zhdanov, Y. A. (1979) *Information Entropy in Organic Chemistry*. Rostov University, Rostov.

Chapter 3

Logarithmic Entropy for a Single Valued Neutrosophic Set

Pratiksha Tiwari*, PhD
Delhi Institute of Advanced Studies, Delhi, India

Abstract

Neutrosophy is a part of philosophy which analyses the origin, nature, and scope of neutralities. A neutrosophic set has three elements represented as: truth, indeterminacy, and falsity memberships. A single valued neutrosophic set (SVNS) is a particular case of neutrosophic set and generalization of fuzzy sets. It deals with uncertainty, imprecise, incomplete, and inconsistent information in a better way. Entropy measure is one of the most important tools to measure uncertainty. In this paper, a logarithmic entropy measure based on Shannon entropy for neutrosophic set has been proposed. Further, a multicriteria decision-making method based on the proposed single valued neutrosophic entropy is established in which criteria values for alternatives are SvNSs.

Introduction

Uncertainty, ambiguity, vagueness, and imprecision are common in human nature and real world problems, fuzzy sets and its generalizations such as intuitionistic fuzzy sets (IFSs) and interval valued intuitionistic fuzzy sets (IvIFSs) are meant to handle such situations. In fuzzy sets belongingness of each element is based on degree of association or membership, whereas IFSs

* Corresponding Author's Email: parth12003@yahoo.co.in.

In: Understanding Information Entropy
Editor: Vijay Kumar
ISBN: 979-8-88697-839-1

and IvIFSs are based on point value or interval for membership and non-membership degrees respectively. Introduction of intervals for membership and non-membership in IvIFSs makes it more flexible and widely applicable to deal with real world problems in recognition of patterns in medical diagnoses, image segmentation and various decision-making problems. Fuzzy set theory (Zadeh, 1965) is a tool that can handle uncertainty and imprecision effortlessly. Interval-valued, intuitionistic, interval-valued intuitionistic fuzzy sets (Zadeh (1975), Atanassov (1986), Atanassov & Gargov (1989)), vague sets (Gau & Buehrer, 1993) and R-fuzzy sets (Yang, Hinde, 2010) are various generalizations of Fuzzy sets (FSs). From all these generalizations IvIFSs and intuitionistic fuzzy sets (IFSs) are two conventional extensions of FSs. IvIFSs are more practical and flexible than IFSs as they are characterized by membership and non-membership degree range instead of real numbers. It makes IvIFSs more useful in dealing with real world complexities which arises due to insufficient information, lack of data, imprecise knowledge, and human nature wherein range is provided instead of real numbers. Distance, entropy, and similarity measures are the central arenas that are investigated by various researchers under intuitionistic and interval-valued fuzzy environment (IFE and IvFE). Neutrosophic theory was recently proposed due to better option where the fuzzy set and fuzzy.

Logic can't express the false membership information and intuitionistic fuzzy set and intuitionistic fuzzy logic is not able to handle the indeterminacy of information. This is due tovagueness and impreciseness of the information that always involve of opposite and neutral information.

Neutrosophic set (NS) developed by Smarandache (1999) is a budding tool for indeterminate data. It can provide a framework for dealing with uncertainty in data including big data. Neutrosophic set generalizes the concept of the classic set, fuzzy set (Zadeh, 1965), interval valued fuzzy set (FS) (Turksen, 1986), intuitionistic fuzzy set (IFS) (Atanassov, 1986), interval valued intuitionistic fuzzy sets (IvIFS) (Atanassov 1999) etc. Due to the special format of these sets, it can handle fuzziness and uncertainty of many practical problems in the real world. NSs are defined by three functions: truth-membership, indeterminacy-membership, and falsity-membership all these functions are represented independently. However, the neutrosophic set generalizes the above-mentioned sets from a philosophical point of view and its functions are real standard or non-standard subsets of]−0, 1+ [, and there is no restriction on the sum of three. Thus, it will be difficult to apply in real scientific and engineering areas. Wang et al., (2010) obtained single-valued neutrosophic set (SvNS), which is a particular case of NS. They also derived set operations and properties for SvNS.

Şahin, & Küçük, (2014) developed a notion of neutrosophic soft set and derived several distance, similarity, and entropy measures for neutrosophic soft sets. Authors used the proposed similarity measures to rank all possible alternatives in multi-criteria decision-making problem. Ye (2014) developed single valued neutrosophic cross entropy which is an extension of the cross entropy of fuzzy sets. Further its application is expressed through multicriteria decision-making problem on SvNS. Broumi & Smarandache (2014) defined cosine similarity based upon Bhattacharya's distance for two interval-valued neutrosophic sets. Majumdar and Samanta (2013), Aydoğdu (2015, 2019) developed similarity and entropy measures for SvNS and demonstrated its application in multi-criteria decision-making. Ye and Cui (2018) proposed exponential entropy measure for simplified neutrosophic set based on Pal and Pal exponential entropy of 1989. They have also investigated the same for single-valued and interval-valued neutrosophic sets with their respective properties. Ngan et al., (2021) done hybridization of SvNS and fuzzy bipolar set and termed it as bipolar neutrosophic set. Further distance, entropy and similarity measures are formulated for bipolar neutrosophic sets. Yang et al., (2019) formed inclusion relation between interval neutrosophic, proposed the axiomatic definitions of the similarity measure and entropy of the interval neutrosophic set. Thereafter proposed a multi-attribute decision-making method using distance and similarity measures to solve decision-making problems and illustrated it with the help of an example.

Various researchers studied different measures such as entropy, cross entropy, distance, and similarity for fuzzy and intuitionistic fuzzy sets that will help in decision-making. The concept of similarity is primarily significant in practically every field. Many methods have been proposed for measuring the degree of similarity between fuzzy and intuitionistic fuzzy sets. But these methods are unsuitable for dealing with the similarity measures of neutrosophic set. Few researchers have dealt with similarity measures for neutrosophic sets. This paper deals with entropy measure for single valued neutrosophic sets and demonstrates its use in multi-attribute decision-making using an illustrative example.

Basic Concepts and Literature Review

Neutrosophic sets emerged as a tool to deal uncertain data. It has potential to become a general framework for uncertainty analysis in data. The neutrosophic set is a part of neutrosophy and generalizes FS, IvFS, IFS, and IvIFS from a philosophical point of view. Smarandache (1999) derived neutrosophic set which is defines as

Definition 1: Let X be a universe of discourse, a neutrosophic set A in X is characterized by a quadruple $\langle x, T_A(x), I_A(x), F_A(x)\rangle$ i.e., (x, truth-membership function, indeterminacy-membership function, falsity-membership function), where $x \in X$. The functions $T_A(x)$: $X \rightarrow]^-0, 1^+[$, $I_A(x)$,: $X \rightarrow]^-0, 1^+[$, and $F_A(x)$: $X \rightarrow]^-0, 1^+[$. Thus, there is no restriction on the sum so $^-0 \leq \sup TA(x) + \sup IA(x) + \sup FA(x) \leq 3^+$.

Wang et al., (2011) proposed a subclass of neutrosophic sets termed as single valued neutrosophic sets which are easier to apply real scientific and engineering problems defined as

Definition 2: Let X universe of discourse, a single valued neutrosophic set (SvNS) A in X is characterized by quadruple $\langle x, T_A(x), I_A(x), F_A(x)\rangle$ i.e., (x, truth-membership function, indeterminacy-membership function, falsity-membership function), where $x \in X$, , $T_A(x), I_A(x), F_A(x) \in [0,1]$ and $0 \leq T_A(x) + I_A(x) + F_A(x) \leq 3$.

Definition 3 (Set operations on single valued neutrosophic sets): Let A and B be two SvNSs defined by quadruple $\langle x, T_A(x), I_A(x), F_A(x)\rangle$ and $\langle x, T_B(x), I_B(x), F_B(x)\rangle$ respectively, where $x \in X$ then set operations are defined as follows:

1. $A \cup B = \{\langle x, T_A(x) \vee T_B(x), I_A(x) \vee I_B(x), F_A(x) \wedge F_B(x)\rangle / x \in X\}$

2. $A \cap B = \{\langle x, T_A(x) \wedge T_B(x), I_A(x) \wedge I_B(x), F_A(x) \vee F_B(x)\rangle / x \in X\}$

3. $\bar{A} = \{\langle x, F_A(x), 1 - I_A(x), T_A(x)\rangle / x \in X\}$

4. $A \subseteq B$ iff $T_A(x) \leq T_B(x)$, $I_A(x) \leq I_B(x)$ and $F_A(x) \geq F_B(x)\ \forall\ x \in X$;

5. $A = B$ iff $T_A(x) = T_B(x)$, $I_A(x) = I_B(x)$ and $F_A(x) = F_B(x)\ \forall\ x \in X$.

Definition 4 (Entropy): Let $N(X)$ be a set of all SvNS and $A \in N(X)$. An entropy on SvNSs is a function $E: N(X) \longrightarrow [0,1]$ which satisfies the following axioms:

1. $E(A) = 0$ if A is a crisp set.
2. $E(A) = 1$ if $\langle T_A(x_i), I_A(x_i), F_A(x_i)\rangle = \langle \frac{1}{2}, \frac{1}{2}, \frac{1}{2}\rangle$ for all $x \in X$.
3. $E(A) \leq E(B)$ if $A \subseteq B$ i.e., if $T_A(x) \leq T_B(x)$, $I_A(x) \leq I_B(x)$ and $F_A(x) \geq F_B(x)$, $\forall\ x \in X$
4. $E(A) = E(A^c)$

Entropy is a major tool to measure uncertainty in data. Zadeh (1965) proposed entropy for fuzzy sets which was based upon the concept of probabilistic entropy of Shannon (1948). Later De Luca and Termini (1968), Pal and Pal (1989) derived entropies for fuzzy sets. Hung & Yang (2006) extended definition for IFSs and IvIFS considering hesitancy degree. Szmidt and Kacprzyk (2000, 2001, 2005a, 2005), Hung and Yang (2006), Ye (2010) and Wang (2012) worked on different entropies under intuitionistic fuzzy environment. Zang et al., (2010), Zhang et al., (2013), Chen et al., (2013), Hu and Li (2013), Zhang et al., (2014), Wei and Zhang (2015), Xie and Lv (2016), Zhao and Xu (2016) introduce different parametric/trigonometric entropies measures for Interval valued intuitionistic fuzzy sets which satisfies the extended form of De Luca-Termini axioms for fuzzy entropy. These researchers also derive relationship between proposed entropy measures and various other information theoretic measures such as distance, similarity, and cross entropy etc. Later in 2014 Majumder and Samanta introduced entropy and similarity measure for single valued neutrosophic sets. Ye and Cui (2018) proposed exponential entropy measure for single valued neutrosophic sets. Aydgdu and Şahinq (2019) derived two distinct measures of entropy for single valued neutroshopic sets and presented its validity through numerical examples in multi attribute decision-making. Wang (2020) proposed entropy and similarity measures for single valued neutrosophic sets with development of its application in multi attribute decision-making. They also proved that the proposed measure is applicable and efficient. Ye (2021) developed trigonometric entropies for simplified neutrosophic sets (SNSs), single-valued neutrosophic sets (SvNSs) and interval-valued neutrosophic sets (IvNSs. Further they developed decision-making approach using ranking using entropy values based upon positive and negative arguments.

In this section some basic definitions of SvNS and existing literature has been discussed. Next section consists of proposed logarithmic entropy.

Logarithmic Entropy Measure

Shannon (1948) introduced the concept termed entropy to measure uncertainty present in information transmission. Rényi (1961) presented generalized logarithmic entropy known as order α entropy. Havrda & Charvat (1967), proposed another generalized measure of entropy called structural β entropy. Zadeh (1968) first introduced fuzzy entropy a measure of fuzzy information based on Shannon's entropy thereafter De Luca and Termini (1972) gave the axiomatic definition of fuzzy entropy and defined logarithmic entropy for fuzzy sets which was based upon Shannon entropy. Verma & Sharma, (2014), proposed generalized intuitionistic fuzzy logarithmic entropy measure of order α considering hesitation index based on Rényi's entropy. Gupta and Tiwari (2016) proposed generalized logarithmic entropy for fuzzy sets. Tiwari (2017) presented logarithmic entropy for intuitionistic fuzzy sets along with weighted correlation measure to identify optimal solution to a decision-making problem.

This section a logarithmic entropy based on Shannon entropy for SvNS is proposed and validity of the same has been proved. The proposed entropy measure is given by equation (1) as follows:

$$E(A) = -\frac{1}{3n}\sum_{i=1}^{n}\begin{bmatrix} T_A(x_i)\log_2\big(T_A(x_i)\big) + (1-T_A(x_i))\log_2\big(1-T_A(x_i)\big) \\ +I_A(x_i)\log_2\big(I_A(x_i)\big) + (1-I_A(x_i))\log_2\big(1-I_A(x_i)\big) \\ +F_A(x_i)\log_2(F_A(x_i)) + (1-F_A(x_i))\log_2\big(1-F_A(x_i)\big) \end{bmatrix}, \text{ for}$$

where $x \in X$, $T_A(x), I_A(x), F_A(x) \in [0,1]$ and $0 \le T_A(x) + I_A(x) + F_A(x) \le 3$ (1)

Next theorem provides the proof that the measure proposed in equation (4.2.1) is a valid measure of interval valued intuitionistic fuzzy entropy.

Theorem 4.2.1: $E(A)$ defined in equation (4.2.1) is a valid measure of interval valued intuitionistic fuzzy entropy.

Proof:

Axiom 1): If A is crisp than $A = [1, 0, 0]$ or $[0, 0, 1]$. Thus $E(A) = 0$.

Axiom 2): Assume that $[T_A(x_i), I_A(x_i), F_A(x_i)]$ then $E(A) = 1$

Conversely, let $E(A) = 1$

$E(A) =$

$$-\frac{1}{3n}\sum_{i=1}^{n}\begin{bmatrix} T_A(x_i)\log_2\big(T_A(x_i)\big) + (1 - T_A(x_i))\log_2\big(1 - T_A(x_i)\big) \\ +I_A(x_i)\log_2\big(I_A(x_i)\big) + (1 - I_A(x_i))\log_2\big(1 - I_A(x_i)\big) \\ +F_A(x_i)\log_2\big(F_A(x_i)\big) + (1 - F_A(x_i))\log_2\big(1 - F_A(x_i)\big) \end{bmatrix} = 1$$

Let $g(c) = -[c\log_2(c) + (1 - c)\log_2(1 - c)]$

where $c \in [0,1]$, we show that $g(c)$ is increasing w.r.t $0 \le c \le \frac{1}{2}$ and decreasing w.r.t. $\frac{1}{2} \le c \le 1$

Equate $\frac{\partial g}{\partial c} = 0$ to identify critical point of g, we get $c = \frac{1}{2}$.

Again, let $c > 0.5$, then

$\frac{dg}{dc} = -[2 + \log_2(c) + \log_2(1 - c)] > 0$. Similarly, it can be proved that when $c < 0.5$, then $g(c)$ is increasing function. Hence $\frac{dg}{dc} > 0$, when $c > 0.5$ and $\frac{\partial g}{\partial c} < 0$, when $c < 0.5$.

Thus $g(c)$ is decreasing w.r.t $c > 0.5$ and increasing for $c < 0.5$, where $c \in [0,1]$. Also $-X\log_2 X - (1 - X)\log_2(1 - X)$ is a well Shannon's function monotonically increasing in the interval [0, 1/2] and monotonically decreasing in [1/2, 1] with a maximum at $X = \frac{1}{2}$.

Hence $E(A)$ function attains its maximum value at $T_A(x_i) = I_A(x_i) = \frac{1}{2} = F_A(x_i)$ Thus $E(A)$ is attains maximum value of 1 at $T_A(x_i) = I_A(x_i) = \frac{1}{2} = F_A(x_i)$.

Axiom 3): To prove that axiom 3) holds good.

Next assume that A is less fuzzy than B. Then $T_A(x_i) \leq T_B(x_i)$, $F_A(x_i) \geq F_B(x_i)$, $I_A(x_i) \leq I_B(x_i)$. By monotonicity $g(c)$, we have $E(A) \leq E(B)$.

Thus, we have established that equation (1) represents a valid measure of entropy for SvNSs. Next section applied the proposed measure of entropy in a multi criteria decision-making problem.

Application of the Proposed Logarithmic Entropy Measure

For multi-attribute decision-making problems with single valued neutrosophic information, this section discusses the application of the proposed logarithmic measure in multi-criteria decision-making based on the concept of entropy. Lesser the value of entropy better the alternative is.

Consider that a fund manager in a wealth management firm is assessing potential investment alternatives $A = \{A_1, A_2, A_3, A_4\}$ to be evaluated based on four attributes risk, growth, socio-political issues and environment impacts represented by $x_1, x_2, x_3,$ and x_4 respectively. Suppose the fund manager comfortable in providing assessment of each alternative based on each attribute in the form of single valued neutrosophic sets and the assessment is as follows:

$$B_1 = \{\langle [0.4,0.4,0.7], [0.1,0.2,0.4], [0.4,0.4,0.4], [0.5,0.5,0.5] \rangle\},$$

$$B_2 = \{\langle [0.8,0.6,0.3], [0.3,0.4,0.4], [0.2,0.8,0.6], [0.6,0.7,0.8] \rangle\},$$

$$B_3 = \{\langle [0.3,0.5,0.4], [0.8,,0.9,0.5], [0.6,0.7,0.9], [0.4,0.5,0.6] \rangle\},$$

$$B_4 = \{\langle [0.7,0.9,0.6], [0.4,0.3,0.7], [0.8,08,05], [0.3,0.2,0.1] \rangle\},$$

The values of $E(B_1) = 0.90816, E(B_2) = 0.865528, E(B_3) = 0.858859, E(B_4) = 0.79757$. Thus, the priority ranking for the four available alternatives based on the proposed entropy measure is $B_4 \succ B_3 \succ$

$B_2 \succ B_1$, corresponding to an ascending order of entropy values where " $\succ$ " corresponds to "superior to." Hence B_4 is the best alternative as it has the least entropy value.

Conclusion

Entropy is a measure of uncertainty devised by Shannon (1948). Shannon proposed a logarithmic entropy for probabilistic measure. Thereafter various generalization of Shannon entropy was done by different researchers in the field of fuzzy, intuitionistic, intuitionistic fuzzy, interval valued, and interval valued intuitionistic fuzzy sets etc. In this paper logarithmic entropy for Single valued neutrosophic set is proposed which is based on Shannon entropy. Further the validation of the same is established and an application of the proposed measure is developed in the field of multi-criteria decision-making.

References

Atanasov, K. T. "Intuitionistic fuzzy sets," *Fuzzy sets and systems*. (1986): 87-96.

Atanassov, K. T. "Interval valued intuitionistic fuzzy sets." In *Intuitionistic fuzzy sets*, pp. 139-177. Physica, Heidelberg, 1999.

Atanassov, K., & Gargov, G. "Interval-valued intuitionistic fuzzy sets," *Fuzzy Sets and Systems*. (1989): 31 (3), 343–349.

Aydoğdu, A. "On similarity and entropy of single valued neutrosophic sets." *General mathematics notes* 29, no. 1 (2015): 67-74.

Aydoğdu, A., and Şahin. R. "*New entropy measures based on neutrosophic set and their applications to multi-criteria decision-making*." Süleyman Demirel Üniversitesi Fen Bilimleri Enstitüsü Dergisi 23, no. 1 (2019): 40-45.

Broumi, S., and Smarandache. F. Cosine similarity measure of interval valued neutrosophic sets. *Infinite Study*, 2014.

Chen, X., Li Yang, Pei Wang, and Wei Yue. "A fuzzy multicriteria group decision-making method with new entropy of interval-valued intuitionistic fuzzy sets." *Journal of Applied Mathematics* 2013 (2013).

De Luca, A., & Termini, S. "A Definition of a Non-probabilistic Entropy in Setting of Fuzzy Sets." *Information and Control* (1972): 20, 301-312.

Gau, W. L. & Buehrer, D. J. "Vague sets," *IEEE Transactions on systems, Man and Cybernetics,* (1993): 23 (2), 610–614.

Gupta, P., and Tiwari, P. "Decision-making approach using weighted coefficient of correlation along with generalized parametric fuzzy entropy measure." *International Journal of Fuzzy System Applications (IJFSA)* 5, no. 3 (2016): 30-41.

Hung, W., & Yang, M. "Fuzzy entropy in intuitionistic fuzzy sets," *International Journal of Intelligent Systems* (2006): Vol. 21, 443-451.

Majumdar, P., and Samanta, S. K. "On similarity and entropy of neutrosophic sets." *Journal of Intelligent & Fuzzy Systems* 26, no. 3 (2014): 1245-1252.

Ngan, R. T., Smarandache, F. T. and Broumi. S. "H-Max Distance Measure of Bipolar Neutrosophic Sets and an Application to Medical Diagnosis." *Neutrosophic Sets and Systems* 45, no. 1 (2021): 29.

Pal, N. R., and Sankar K. Pal. "Object-background segmentation using new definitions of entropy." *IEE Proceedings E-Computers and Digital Techniques* 136, no. 4 (1989): 284-295.

Şahin, R., and Ahmet Küçük. "On similarity and entropy of neutrosophic soft sets." *Journal of intelligent & fuzzy systems* 27, no. 5 (2014): 2417-2430.

Shannon, C. E. "A mathematical theory of communication." *The Bell system technical journal* 27, no. 3 (1948): 379-423.

Smarandache, F. "A unifying field in Logics: Neutrosophic Logic." In *Philosophy,* pp. 1-141. American Research Press, 1999.

Szmidt, E., and Janusz Kacprzyk. "A new measure of entropy and its connection with a similarity measure for intuitionistic fuzzy sets." In *EUSFLAT Conf.,* pp. 461-466. 2005a.

Szmidt, E., and Janusz Kacprzyk. "Distances between intuitionistic fuzzy sets." *Fuzzy sets and systems* 114, no. 3 (2000): 505-518.

Szmidt, E., and Janusz Kacprzyk. "Entropy for intuitionistic fuzzy sets." *Fuzzy sets and systems* 118, no. 3 (2001): 467-477.

Szmidt, E., and Janusz Kacprzyk. "New measures of entropy for intuitionistic fuzzy sets." In *Ninth Int Conf IFSs Sofia*, vol. 11, no. 2, pp. 12-20. 2005b.

Tiwari, P. "Generalised intuitionistic fuzzy entropy and weighted correlation with application in multi-attributes decision-making." *International Journal of Fuzzy Computation and Modelling* 2, no. 3 (2017): 261-274.

Turksen, I. B. "Interval valued fuzzy sets based on normal forms." *Fuzzy sets and systems* 20, no. 2 (1986): 191-210.

Verma, R., and Sharma, B. D. "Exponential entropy on intuitionistic fuzzy sets." *Kybernetika* 49, no. 1 (2013): 114-127.

Wang, H., Y. Zhang, R. Sunderraman, and F. Smarandache. "Single valued neutrosophic sets, Fuzzy Sets." *Rough Sets and Multivalued Operations and Applications* 3, no. 1 (2011): 33-39.

Wang, H. "Fuzzy entropy and similarity measure for intuitionistic fuzzy sets." *Advances in Biomedical Engineering* 10 (2012): 84.

Wei, C., and Yuzhong Zhang. "Entropy measures for interval-valued intuitionistic fuzzy sets and their application in group decision-making." *Mathematical Problems in Engineering* (2015).

Xie, X. J., and Xiao-xia Lv. "Improved interval-valued intuitionistic fuzzy entropy and its applications in multi-attribute decision-making problems." In *Fuzzy Systems & Operations Research and Management*, pp. 201-211. Springer, Cham, 2016.

Yang, H. Xiaoman Wang, and Keyun Qin. "New similarity and entropy measures of interval neutrosophic sets with applications in multi-attribute decision-making." *Symmetry* 11, no. 3 (2019): 370.

Yang, Y. J. & Hinde, C., "A new extension of fuzzy sets using rough sets: R-fuzzy sets," *Information Sciences* (2010), 180, 354–365.

Ye, J., Cui W. "Exponential Entropy for Simplified Neutrosophic Sets and Its Application in Decision-making." *Entropy* (Basel). 2018 May 10;20(5):357. doi: https://doi.org/10.3390/e20050357.

Ye, J. and Wenhua Cui. "Exponential entropy for simplified neutrosophic sets and its application in decision-making." *Entropy* 20, no. 5 (2018): 357.

Ye, J. "Entropy measures of simplified neutrosophic sets and their decision-making approach with positive and negative arguments." *Journal of Management Analytics* 8, no. 2 (2021): 252-266.

Ye, J. "Single valued neutrosophic cross-entropy for multicriteria decision-making problems." *Applied Mathematical Modelling* 38, no. 3 (2014): 1170-1175.

Ye, J. "Two effective measures of intuitionistic fuzzy entropy." *Computing* 87, no. 1 (2010): 55-62.

Zadeh, L. A, George J. K., and Yuan, B. Fuzzy sets, fuzzy logic, and fuzzy systems: selected papers. Vol. 6. *World Scientific,* 1996.

Zadeh, L. A., "Fuzzy sets," *Information and Control* (1965): 8 (3), 338–356.

Zadeh, L. A., The concept of a linguistic variable and its application to approximate reasoning-I, *Information Science*, 8 (1975) 199–249.

Zhang, H. "Entropy for intuitionistic fuzzy sets based on distance and intuitionistic index." *International Journal of Uncertainty, Fuzziness and Knowledge-Based Systems* 21, no. 01 (2013): 139-155.

Zhang, Q. S., Shengyi Jiang, Baoguo Jia, and Shihua Luo. "Some information measures for interval-valued intuitionistic fuzzy sets." *Information sciences* 180, no. 24 (2010): 5130-5145.

Zhao, N., and Zeshui Xu. "Entropy measures for interval-valued intuitionistic fuzzy information from a comparative perspective and their application to decision-making." *Informatica* 27, no. 1 (2016): 203-229.

Chapter 4

A Hybrid Intuitionistic Fuzzy Information Based TOPSIS Method with Applications in Multi Criteria Decision-Making

H. D. Arora[1]
Anjali Naithani[1]
and Rajender Kumar[2,*]
[1]Department of Applied Mathematics, Amity Institute of Applied Sciences, Amity University, Noida, India
[2]Department of Mechanical Engineering, Faculty of Engineering & Technology, Manav Rachna International Institute of Research & Studies, Faridabad, Haryana

Abstract

Choosing a university that provides quality education and meets all of the requirements for a specific student is a difficult task. Most people are hesitant to share information that affects their decision, but it is critical to consider all of the parameters associated with the selection of a specific university. In this study, the TOPSIS approach based on intuitionistic fuzzy sets is used to make the university selection decision. Multiple academic and non-academic factors have been considered here, which are difficult to consider and even impossible to define based on the associated facts. The study reports and ranks the parameters associated with the selection of a university/instution.

Keywords: selection of university, education, decision-making, TOPSIS, multi-criteria-decision-making approach (MCDM)

[*] Corresponding Author's Email: rajender6291@gmail.com.

In: Understanding Information Entropy
Editor: Vijay Kumar
ISBN: 979-8-88697-839-1

Introduction

Education in general, is the process that facilitates learning among humans. It is also known as the process of acquisition of the learning's. One can learn from the skillset, value system, belief, and habit, etc. The various methods of knowledge transfer among humans are teaching, training, storytelling, discussion, and directed research, etc., (Poock et al., 2001). The mode of delivery is one of the important attributes while looking for the output of the education system (Soutar and Turner, 2002).

The education sector's uniqueness keeps it a priority in the nation's development and growth by contributing to human welfare development (Briggs and Wilson, 2007). The sector has the intention to serve all kinds of human beings. Globally, all nations are attempting to build or develop education as one of the fastest growing sectors because it has a direct impact on a country's living and economic status. The facts and facets of today's education sector confirm the need for radical changes, not only in curriculums but also in lecture delivery methods (Briggs, 2006). Even though, in the past decade, several techniques and tools were developed to carry the education sector in the direction of next-generation educational contexts, Many of them are related to the courses (design and development); admission and exit programs; and selection of the best university or institution, etc.

As of today, almost 50–70% of the educational institutions are run by private companies. It is a well-known fact that the name and reputation of the institution are extremely important in the lives of students who graduate from it. The names of the institutions where they receive long-term education and help with their job profile, lifestyle, salary, and so on (Warwick and Mansfield, 2004). That is why it is pertinent to address the issue, which is very crucial, i.e., the decision regarding the selection of an institution or university. Entrants to any institution or university go through a life-changing transition that aids in the development of a career based on the present and future aspects. Almost all entrants are gathering information about the institution or university on various parameters before making a decision to enroll. In actual practice, it seems to be difficult to make decisions regarding the selection of the best university based on various parameters (Simoes and Soares, 2010). In this regard, the MCDM approach plays an important role because it considers several parameters and criteria. Here, it is an exaggeration to say that any small error in the consideration of a parameter or criteria can influence the decision. So, it becomes important to select the parameter carefully, and then the approach is used to model the final solution (Sojkin et al., 2011).

The existing literature reports that there are two important categories for the selection of institutions and universities, i.e., academic and non-academic categories. In the academic category, there are factors such as information about an institution's or university's reputation, faculty members involved in the teaching-learning process, the number of courses offered to students, the quality and variety of programmes offered, and social criteria such as future options and prospects, life within the university, other student groups, transparency in the university's information dissemination, family beliefs and influence, and so on (Pampaloni, 2010; Kumar et al., 2016). Non-academic parameters include geographical location, infrastructural adequacy, languages used in the teaching-learning process, and the need for special facilities, particularly for disabled students. The other parameters are the course fee, which includes tuition and administrative charges, various scholarship offers, and the cost of living within or outside the campus (Singh et al., 2019; Kumar et al., 2021).

The decision on which university to attend is made in the current study. For this purpose, the selection parameters, such as academic excellence, administrative support (non-academic), sociological and psychological factors, and many more, are studied from the published literature. Though there are studies available on the selection of a university, the present study will address the gaps in earlier studies using the fuzzy-TOPSIS approach. This is the intuitionistic fuzzy method in which the various parameters identified from the literature and other sources are then systematically related to the criteria (also identified from the literature and other sources) and the individual parameter Engül and colleagues (2015). This will help in identifying the rank of an individual parameter. The parameters ranked at position 1 or 2 will be addressed in priority, and the process will continue until all parameters are addressed.

2. Literature Review

Earlier, the preferred mode of education was offline and based on set procedures defined by various regulatory bodies. The offline systems seek the presence of both the knowledge provider and the learner in a common area such as a classroom, meeting hall, etc., Taylor et al., 2008; Adams and Evenland, 2007). Aspirants are increasingly preferring or appreciating online education. Due to this, the higher education sector's contribution to the nation's economy has increased drastically. Still, there is an important concern before the educational institutions or universities, i.e., the selection of the institution or university by the maximum number of aspirants, which will further contribute to the profit maximisation of the institution or university. Most educational institutions, particularly those run by private players, are now having difficulty enrolling students (Tomte et al., 2015). The institutions are not selected by the students because they do not fulfil the criteria for selection. Most of the students are considering the institution selection as a decision pertaining to their future and wish to ensure its improvement. It helps them get good perks and a secure life in terms of economic considerations.

Literature published on the institution/university selection reveals that statistical methods are being used for optimising the selection parameter (Reese, 2015). The geo-graphical positioning of an institution, the approachability of the campus, the courses offered by the university, the selection criteria, and the faculties were the most commonly reported concerns for university or institution selection in previous studies. Even so, the studies also reported that the availability of visas and the cost of living while looking for an institution or university abroad were the main concerns (Shapiro et al., 2017; Damary et al., 2017).

For getting the solution, the literature reports that the deployment of multi-criteria decision-making approaches is one of the best alternatives (Hwang and Yoon, 1981). The approaches provide solutions by identifying the most influential or contributing factor in the decision. This is done based on the approach considered in the study, and in some approaches, like Entropy, the criteria are also considered (Yazdani et al., 2015; Zhang et al., 2011). All the possible factors are ranked in such a manner that the factor ranked first is the most important one, and so on.

While using the MCDM approaches, the presence of error means the vague data inputs for the factors can impact the decisions (Szmidt and Kacprzyk, 2000, 2004). Fuzzy methods are being used by researchers and practitioners to avoid such errors (Liu and Wang, 2007; Xu and Yegar, 2008). The TOPSIS approach is a widely used approach for the selection of parameters and factors in a fuzzy environment (Chen and Li, 2010; Li et al., 2011). This is the classical method in which the alternatives are categorised into positive ideal solutions (closer) and negative ideal solutions (far) (Xu et al., 2013). The TOPSIS approach is very simple and rational in its use. In addition, this approach has the computational efficiency and ability to measure the relative performance of each alternative (Chakraborty and Chatterjee, 2013). In the TOPSIS approach, the intuitive and clear logic regarding the parameters is considered to represent the rationale of human choice (Kumar et al., 2018; Chen, 2019). It helps the decision-maker in constructing the problem and devising the solution in such a manner that comparison among factors can be done easily to define the individual parameter rank. The input data in this case is the classic TOPSIS method for solving problems, where all decision data is known and represented by crisp numbers. Most of the real-world problems, however, have a more complex structure with uncertainty, lack of information, or vagueness (Elikbilek and Tüysüz, 2020; Sindhwani et al., 2021). The criteria's importance and the alternative's impact on the criteria given by decision makers are difficult to represent in crisp data, so IFS are appropriate in a multi-criteria decision-making environment.

3. Methodology

The current study takes into account the admission purpose selection. The parameters pertaining to the selection of an institution or university are identified in the literature. Further, the MCDM approach named TOPSIS is used to rank the various parameters based on their respective priority sets (Rouyendegh et al., 2019).

3.1. Preliminaries

The first sets of fuzzy were proposed in 1965 to model vagueness, and this was an advancement in the classical fuzzy set theory (Zadeh, 1965). The fuzzy sets theory was extended for decision-making accuracy and addressed the issue of vagueness (Atanassov, 1986). In the fuzzy set, the intuitionistic fuzzy set is denoted by *'I'* and the universe is denoted by *'X'*. The intuitionistic fuzzy value is calculated with the following equation:

$$I = \{< \mathrm{x}, \xi_I(x), \rho_I(x) >: \forall \mathrm{x} \in \mathrm{X}\}$$

where, the functions $\xi_I(x)$ *and* $\rho_I(x): X \to [0,1]$

The value of X represents the degree of membership and the degree of non-membership of an element $x \in I \subset X$ respectively.

$$\varphi_I(x) = 1 - \xi_I(x) - \rho_I(x)$$

where, degree of uncertainty is calculated with condition i.e., $0 \leq \boldsymbol{\xi_I(x) + \rho_I(x) \leq 1}$. The parameter $\boldsymbol{\varphi_I(x)}$ is hesitation of degree whether x belongs to I or not. Low value of parameter $\boldsymbol{\varphi_I(x)}$ indicates the information about x is more certain whereas the high $\boldsymbol{\varphi_I(x)}$ value implies the uncertainty in information about x, (Shu et al., 2006) when $\boldsymbol{\xi_I(x) = 1 - \rho_I(x)}\ \forall \boldsymbol{x}$, it becomes fuzzy set. For IFS U and V of set X, multiplication operator is given by:

$$U \otimes V = \{\xi_I(x) \bullet \xi_J(x), \rho_I(x) + \rho_J(x) - \rho_I(x) \bullet \rho_J(x)\ |\mathrm{x} \in \mathrm{X}\} \tag{1}$$

(Atanassov, 1986, 1999)

3.2. An Algorithm for Intuitionistic Fuzzy TOPSIS

Let $I = \{I_1, I_2, \dots\dots. I_m$ be the set of m alternatives and $X = \{X_1, X_2, \dots\dots. X_n\}$ be set of n criteria. Steps involved in intuitionistic fuzzy TOPSIS method are following: (Xu et al., 2013; Alidrisi and Al-Sasi, 2017).

Step-1-Decision makers weight evaluation: The weight for each parameter is identified with the help of linguistic terms. These tems are expressed in intuitionistic fuzzy numbers for determining the importance of 'r' decision maker. Let $Q_{k=}[\xi_k, \rho_k, \varphi_k]$ be an IF number to rate the kth decision maker (Grzegorzewski, 2004; Attansov et al., 2005). Thus, the kth decision maker's weight is evaluated as:

$$\gamma_k = \frac{(\xi_k + \varphi_k(\frac{\xi_k}{\xi_k + \rho_k}))}{\sum_{k=1}^{l}(\xi_k + \varphi_k(\frac{\xi_k}{\xi_k + \rho_k}))} \tag{2}$$

and

$$\sum_{k=1}^{l} \gamma_k = 1$$

Step-2-Aggregated IFS Matrix - DM's Opinion: Let $P^{(k)} = (p_{ij}^{(k)})_{m\times n}$ is an IF decision matrix, $\gamma = \{\gamma_1, \gamma_2, \ldots\ldots \gamma_l\}$ denotes each decision maker's weight and $\sum_{k=1}^{l} \gamma_k = 1, \gamma_k \in [0,1]$.

In multi criteria decision-making method, all the individual opinions should be intertwined into a gathered conclusion to develop an aggregated intuitionistic fuzzy decision matrix (Wang and Xin, 2005; Xu and ager, 2006). The IFWA Operator is utilized to achieve that $P = (p_{ij})_{m\times n}$

$$P = \begin{bmatrix} (\xi_{A_1(x_1)}, \rho_{A_1(x_1)}, \varphi_{A_1(x_1)}) & (\xi_{A_1(x_2)}, \rho_{A_1(x_2)}, \varphi_{A_1(x_1)}) & \cdots & (\xi_{A_1(x_n)}, \rho_{A_1(x_n)}, \varphi_{A_1(x_n)}) \\ (\xi_{A_2(x_1)}, \rho_{A_2(x_1)}, \varphi_{A_2(x_1)}) & (\xi_{A_2(x_2)}, \rho_{A_2(x_2)}, \varphi_{A_2(x_2)}) & \cdots & (\xi_{A_2(x_n)}, \rho_{A_2(x_n)}, \varphi_{A_2(x_n)}) \\ \vdots & \vdots & \ddots & \vdots \\ (\xi_{A_m(x_1)}, \rho_{A_m(x_1)}, \varphi_{A_m(x_1)}) & (\xi_{A_m(x_2)}, \rho_{A_m(x_2)}, \varphi_{A_m(x_2)}) & \cdots & (\xi_{A_m(x_n)}, \rho_{A_m(x_n)}, \varphi_{A_m(x_n)}) \end{bmatrix} \quad (3)$$

where,

$$\begin{aligned} p_{ij} &= IFWA_{\psi}\left(p_{ij}^{(1)}, p_{ij}^{(2)}, \ldots\ldots\ldots, p_{ij}^{(l)}\right) \\ &= \gamma_1 p_{ij}^{(1)} \oplus \gamma_2 p_{ij}^{(2)} \oplus \gamma_3 p_{ij}^{(3)} \oplus \ldots\ldots\ldots\ldots \gamma_l p_{ij}^{(l)} \\ &= [1 - \prod_{k=1}^{l}\left(1 - \xi_{ij}^{(k)}\right)^{\gamma_k}, \prod_{k=1}^{l}\left(\rho_{ij}^{(k)}\right)^{\gamma_k}, \prod_{k=1}^{l}\left(1 - \xi_{ij}^{(k)}\right)^{\gamma_k} - \prod_{k=1}^{l}(\rho_{ij}^{(k)})^{\gamma_k}] \end{aligned} \quad (4)$$

Here

$$p_{ij} = \left(\xi_{I_i}(x_j) \cdot \rho_{I_i}(x_j), \varphi_{I_i}(x_j)\right) (i = 1, 2, \ldots\ldots\ldots, m\,; j = 1, 2, \ldots\ldots\ldots, n).$$

$$P\begin{bmatrix} p_{11} & p_{12} & \cdots & p_{1m} \\ p_{21} & p_{22} & \cdots & p_{2m} \\ \cdots & \cdots & \cdots & \cdots \\ p_{n1} & p_{n2} & \cdots & p_{nm} \end{bmatrix} \quad (5)$$

(Xu, 2007)

Step-3-Weight of Criteria Using Entropy: In this step weight vector $W = (w_1^g, w_2^g, \ldots\ldots, w_n^g)$ is obtained using intuitionistic fuzzy entropy measure. (Zeleny, 1976; Conard, 2000; Chen and Li, 2010) where $w_i^g \geq 0$ and $\sum_{i=1}^{n} w_i^g = 1$. The entropy values of each IFS is evaluated, the value in the decision matrix P is expressed into an entropy value H_{ij} as shown in Equation 6 where H_{ij} is intuitionistic fuzzy entropy:

$$P = \begin{bmatrix} H_{11} & H_{12} & H_{13} & \cdots & H_{1m} \\ H_{21} & H_{22} & H_{23} & \cdots & H_{2m} \\ H_{31} & H_{32} & H_{33} & \cdots & H_{3m} \\ \cdots & \cdots & \cdots & \ddots & \vdots \\ H_{n1} & H_{n2} & H_{n3} & \cdots & H_{nm} \end{bmatrix} \tag{6}$$

The intuitionistic fuzzy entropy values are normalized in the decision matrix using Equation (7).

$$g_{ij} = \frac{H_{i1}}{\max(H_{i1})}, \frac{H_{i2}}{\max(H_{i2})}, , , , \frac{H_{ij}}{\max(H_{ij})} \tag{7}$$

where, $i = 1, 2, \ldots, l; j = 1, 2, \ldots, m$

h_{ij} defines the normalized value. The decision matrix as obtained after normalization is as follows:

$$P = \begin{bmatrix} g_{11} & g_{12} & g_{13} & \cdots & g_{1m} \\ g_{21} & g_{22} & g_{23} & \cdots & g_{2m} \\ g_{31} & g_{32} & g_{33} & \cdots & g_{3m} \\ \cdots & \cdots & \cdots & \ddots & \vdots \\ g_{n1} & g_{n2} & g_{n3} & \cdots & g_{lm} \end{bmatrix}$$

Objective weights w_j^g of criteria x_j thus obtained as:

$$w_j^g = \frac{1}{n-T} \times (1 - e_j) \tag{8}$$

where $e_j = \sum_{i=1}^{m} g_{ij}$ and $Q = \sum_{j=1}^{n} e_j$

Step-4-Weighted Aggregated IFS Matrix - DM's Opinion: Weighted aggregated intuitionistic fuzzy decision matrix are constructed utilizing following definition: (Zadeh, 1965, 1973).

$$P \otimes W = \{< x, \xi_{I_i}(x) \cdot \xi_W(x), \rho_{I_i}(x) + \rho_W(x) - \rho_{I_i}(x) \cdot \rho_W(x) > | x \in X \}$$
$$\text{and } \varphi_{I_i \cdot W}(x) = 1 - \rho_{I_i}(x) - \rho_W(x) - \xi_{I_i}(x) \cdot \xi_W(x) + \rho_{I_i}(x) \cdot \rho_W(x) \quad (9)$$

$$P' = \begin{bmatrix} p'_{11} & p'_{12} & p'_{13} & \cdots & p'_{1j} \\ p'_{21} & p'_{22} & p'_{23} & \cdots & p'_{2j} \\ p'_{31} & p'_{32} & p'_{33} & \cdots & p'_{3j} \\ \cdots & \cdots & \cdots & \ddots & \vdots \\ p'_{i1} & p'_{i2} & p'_{i3} & \cdots & p'_{ij} \end{bmatrix}$$

$$P' = \begin{bmatrix} (\xi_{A_1W(x_1)}, \rho_{A_1W(x_1)}, \varphi_{A_1W(x_1)}) & (\xi_{A_1W(x_2)}, \rho_{A_1W(x_2)}, \varphi_{A_1W(x_2)}) & \cdots & (\xi_{A_1W(x_n)}, \rho_{A_1W(x_n)}, \varphi_{A_1W(x_n)}) \\ (\xi_{A_2W(x_1)}, \rho_{A_2W(x_1)}, \varphi_{A_2W(x_1)}) & (\xi_{A_2W(x_2)}, \rho_{A_2W(x_2)}, \varphi_{A_2W(x_2)}) & \cdots & (\xi_{A_2W(x_n)}, \rho_{A_2W(x_n)}, \varphi_{A_2W(x_n)}) \\ \vdots & \vdots & \ddots & \vdots \\ (\xi_{A_mW(x_1)}, \rho_{A_mW(x_1)}, \varphi_{A_mW(x_1)}) & (\xi_{A_mW(x_2)}, \rho_{A_mW(x_2)}, \varphi_{A_mW(x_2)}) & \cdots & (\xi_{A_mW(x_n)}, \rho_{A_mW(x_n)}, \varphi_{A_mW(x_n)}) \end{bmatrix} \quad (10)$$

where, $p'_{ij} = (\xi'_{ij}, \rho'_{ij}, \varphi'_{ij}) = (\xi_{I_iW}(x_j), \rho_{I_iW}(x_j), \varphi_{I_iW}(x_j))$

Step-5-Intuitionistic Fuzzy Set Deviations: Let K_1 be benefit criteria and K_2 be the cost criteria, α^+ is IFPIS - intuitionistic fuzzy positive ideal solution and α^- is IFNIS - intuitionistic fuzzy negative ideal solution:

$$\alpha^+ = (\xi_{A^*W}(x_j), (\rho_{A^*W}(x_j)) \text{ and } \alpha^- = (\xi_{A^-W}(x_j), (\rho_{A^-W}(x_j))$$

where

$$\xi_{I^*W}(x_j) = \left(\left(\max_i \xi_{I_iW}(x_j) \,|\, j \in J_1 \right), \left(\min_i \xi_{I_iW}(x_j) \,|\, j \in J_2 \right) \right)$$

$$\rho_{I^*W}(x_j) = \left(\left(\min_i \rho_{I_iW}(x_j) \,|\, j \in J_1 \right), \left(\max_i \rho_{I_iW}(x_j) \,|\, j \in J_2 \right) \right)$$

$$\xi_{I^-W}(x_j) = \left(\left(\min_i \xi_{I_iW}(x_j) \,|\, j \in J_1 \right), \left(\max_i \xi_{I_iW}(x_j) \,|\, j \in J_2 \right) \right)$$

$$\rho_{I^-W}(x_j) = \left(\left(\max_i \rho_{I_iW}(x_j) \,|\, j \in J_1 \right), \left(\min_i \rho_{I_iW}(x_j) \,|\, j \in J_2 \right) \right) \quad (11)$$

Step-6-Separation Measures: The separation measures, $\theta_{IFS}(\alpha_i, \alpha_i^+)$ and $\theta_{IFS}(\alpha_i, \alpha_i^-)$, of each alternative from IFPIS and IFNIS are calculated using following distance measure,

$$\theta_{IFS}(\alpha_i, \alpha_i^+) = \sum_{j=1}^{n} \max\left(\left|\xi_{I_iW}(x_j) - \xi_{I^+W}(x_j)\right|, \left|\rho_{I_iW}(x_j) - \rho_{I^+W}(x_j)\right|\right)$$

$$\theta_{IFS}(\alpha_i, \alpha_i^-) = \sum_{j=1}^{n} \max\left(\left|\xi_{I_iW}(x_j) - \xi_{I^-W}(x_j)\right|, \left|\rho_{I_iW}(x_j) - \rho_{I^-W}(x_j)\right|\right) \tag{12}$$

Step-7-Closeness Coefficient: The relative closeness coefficient with respect to the IFPIS α_i^+ of an alternative α_i is given as follows:

$$C_{i^*} = \frac{\theta_{IFS}(\alpha_i, \alpha_i^-)}{\theta_{IFS}(\alpha_i, \alpha_i^+) + \theta_{IFS}(\alpha_i, \alpha_i^-)} \tag{13}$$

where $0 \leq C_{i^*} \leq 1, i = 1, 2, 3, \ldots, l$

Step-8-Alternative's Rank: The alternatives are assigned ranks in decreasing order after determining the C_{i^*}.

4. Case Study on Criteria Based University/Institution Selection

Any individual (whether seeking admission for himself/herself or for a relative) has numerous questions regarding the institution or university selection. Due to this, most people define their parameters and evaluate them before making a decision regarding a particular university (Broekemier and Seshadiri, 2000). They gathered the information pertaining to those parameters and evaluated them based on various criteria. The process of gathering information and evaluating it for a decision is cumbersome and requires resources. In addition, inadequate information regarding the parameters or criteria can have an impact on the student's carer (Conard and

Conard, 2000). Using advanced methods, it becomes necessary to address the issue of selection parameters and their impact on or contribution to the selection criteria. The two main criteria for the selection of an institution or university are academic and non-academic. The present study is based on parameter selection and ranking them by the TOPSIS approach. The work will provide information to various aspirants seeking university admission. For the work, the ten selection criteria were identified through literature and are listed below:

1) Faculty/Research Facility
2) University Reputation/Prestige
3) Programs offered
4) quality in executing the programme
5) Location of the university
6) Infrastructural adequacy
7) Future career aspects and opportunities
8) Tuition Fee
9) Living expense
10) Scholarships

The identified criteria are reported as key points to assist aspirants in making an informed decision without hesitation. Now, the model is being developed with the help of a multi-criteria decision-model approach to select the most appropriate factors to take into consideration while choosing or selecting an institution or university. The steps of the TOPSIS approach are as follows:

Step-1-Decision makers weight evaluation: Before moving further, it is important to define the significance of decision-makers. Here, the intrapersonal capabilities of an individual decision-maker contribute a lot. It is a well-known fact about humans that their ability to judge or evaluate is critical and varies from person to person. For this reason, the fuzzy intuitionistic method is used. It aids in obtaining information from experts in such a way that the data must be contained within the IFS. The expert's significance is estimated using the eq. (5) mentioned earlier.

In addition, the linguistic terms are used in the present study reported in Table 1.

Table 1. Lingustic terms

Linguistics Terms	Very-Unimportant	Unimportant	Medium	Important	Very-Important
IFN's	0.3	0.5	0.5	0.8	0.9
	0.65	0.45	0.15	0.45	0.1

Further, the resultant weights estimated and reported in table-2. are represented in Table 2.

Table 2. Weight of decision maker

Decision Makers	Linguistics Term	Weight (λ)
I	Very Important	0.397
II	Medium	0.232
III	Important	0.371

Step-2-Aggregated IFS Matrix based on DM's Opinions: Linguistic terms represented in Table 3 with notations Extremely Good(L_1), Very Very Good (L_2), Very Good (L_3), Good (L_4), Medium Good (L_5), Fair (L_6), Medium Bad (L_7), Bad (L_8), Very Bad (L_9), Very Very Bad (L_{10}) are used to construct the opinion of decision maker based on all criteria for each alternative expressed in Table 4, using Equation (7) aggregated decision matrix is obtained as represented in Table 5.

Table 3. Linguistic terms

Linguistic Terms	L_1	L_2	L_3	L_4	L_5	L_6	L_7	L_8	L_9	L_{10}
IFNs	0.95	0.85	0.75	0.7	0.65	0.55	0.4	0.25	0.1	0.1
	0.05	0.1	0.15	0.2	0.25	0.35	0.5	0.65	0.75	0.9

Table 4. Linguistic terms

	Criteria X_1				Criteria X_2				Criteria X_3				Criteria X_5				Criteria X_4			
	A_1	A_2	A_3	A_4	A_1	A_2	A_3	A_4	A_1	A_2	A_3	A_4	A_1	A_2	A_3	A_4	A_1	A_2	A_3	A_4
DM_1	D	VVG	MG	F	MG	VG	VG	G	VG	VG	VG	G	F	VG	MG	VG	G	VVG	G	VG
DM_2	VG	VG	G	MG	G	G	VG	MG	G	VG	VG	G	MG	VG	VG	G	VG	G	F	G
DM_3	G	VG	G	MB	MG	VG	MG	G	VG	G	G	MG	F	VVG	VG	MG	G	VG	F	G

	Criteria X_6				Criteria X_7				Criteria X_8				Criteria X_9				Criteria X_{10}			
	A_1	A_2	A_3	A_4	A_1	A_2	A_3	A_4	A_1	A_2	A_3	A_4	A_1	A_2	A_3	A_4	A_1	A_2	A_3	A_4
DM_1	MG	VG	MG	F	G	VVG	VG	MB	G	VVG	VG	G	VG	VVG	MG	F	G	VG	MG	MB
DM_2	VG	G	MG	MB	VG	MG	G	MB	MG	VVG	MG	VG	MG	G	G	MB	MG	G	MG	MB
DM_3	G	VVG	G	G	VG	VG	MG	G	VG	VG	VG	VG	VG	VVG	G	G	G	VG	MG	MG

Table 5. Aggregated IF decision matrix

Criteria	Alternative (A_1)	Alternative (A_2)	Alternative (A_3)	Alternative (A_4)
X_1	(0.712, 0.187)	(0.796, .0128)	(0.681, 0.219)	(0.528, 0.37)
X_2	(0.662, 0.237)	(0.739, 0.16)	(0.717, 0.181)	(0.689, 0.211)
X_3	(0.739, 0.16)	(0.733, 0.167)	(0.733, 0.167)	(0.682, 0.217)
X_4	(0.575, 0.324)	(0.793, 0.129)	(0.714, 0.184)	(0.705, 0.194)
X_5	(0.712, 0.187)	(0.787, 0.137)	(0.617, 0.28)	(0.721, 0.178)
X_6	(0.694, 0.204)	(0.784, 0.138)	(0.669, 0.23)	(0.586, 0.309)
X_7	(0.731, 0.17)	(0.779, 0.155)	(0.705, 0.2)	(0.536, 0.405)
X_8	(0.709, 0.194)	(0.819, 0.119)	(0.73, 0.174)	(0.731, 0.17)
X_9	(0.73, 0.174)	(0.824, 0.124)	(0.681, 0.22)	(0.586, 0.339)
X_{10}	(0.689, 0.212)	(0.739, 0.162)	(0.65, 0.25)	(0.509, 0.419)

Step-3-Weight of Criteria Using Entropy: Decision makers evaluated the ten criteria and their opinions were aggregated to estimate the weights of each criteria using Equation (6), the resultant weighs are represented in Table 6,

Table 6. Weights of criteria

Criteria	X_1	X_2	X_3	X_4	X_5
Weight	0.0564	0.10053	0.09935	0.0643	0.075
Criteria	X_6	X_7	X_8	X_9	X_{10}
Weight	0.0754	0.05189	0.10057	0.0634	0.0588

Step-4-Weighted Aggregated IFS Matrix - DM's Opinion: In the step-4, the relation represented in equation-7 is estimated to provide the exact information regarding weighted aggregated intuitionistic fuzzy decision matrix. The table-7 reports the weighted aggregated.

Table 7. Weighted aggregated IF decision matrix

Criteria	A_1	A_2	A_3	A_4
X_1	(0.068, 0.91	(0.086, 0.890)	(0.062, 0.920)	(0.042, 0.950)
X_2	(0.103, 0.865)	(0.126, 0.832)	(0.119, 0.842)	(0.111, 0.855)
X_3	(0.125, 0.834)	(0.123, 0.837)	(0.123, 0.837)	(0.108, 0.860)
X_4	(0.054, 0.930)	(0.097, 0.877)	(0.077, 0.897)	(0.076, 0.900)
X_5	(0.089, 0.882)	(0.109, 0.862)	(0.069, 0.910)	(0.091, 0.880)
X_6	(0.085, 0.887)	(0.109, 0.861)	(0.080, 0.895)	(0.064, 0.915)
X_7	(0.071, 0.905)	(0.082, 0.900)	(0.067, 0.913)	(0.042, 0.950)
X_8	(0.067, 0.912)	(0.092, 0.887)	(0.071, 0.906)	(0.072, 0.905)
X_9	(0.071, 0.906)	(0.093, 0.888)	(0.062, 0.918)	(0.048, 0.941)
X_{10}	(0.064, 0.916)	(0.073, 0.902)	(0.057, 0.925)	(0.039, 0.952)

Step-5-Intuitionistic Fuzzy Set Deviations: The step 5 reports that the value of Intuitionistic fuzzy ideal solution in terms of positive as well as negative. The equation-9 is used to calculate the values for IFPIS and IFNIS. The evaluated values are furnished in table 8.

Table 8. IFPIS and IFNIS

Criteria	α^+	α^-	Criteria	α^+	α^-
X_1	0.0857	0.0415	X_6	0.1091	0.0643
	0.89053	0.94547		0.86137	0.91531
	0.02372	0.01307		0.02958	0.0204
X_2	0.1263	0.1033	X_7	0.0816	0.0424
	0.8317	0.8653		0.9002	0.9503
	0.042	0.031		0.018	0.007
X_3	0.125	0.108	X_8	0.092	0.067
	0.834	0.859		0.887	0.912
	0.0415	0.0332		0.0212	0.0211
X4	0.096	0.054	X_9	0.093	0.049
	0.8765	0.9301		0.8889	0.9408
	0.027	0.016		0.018	0.011
X_5	0.109	0.069	X_{10}	0.073	0.039
	0.8616	0.909		0.9024	0.9521
	0.029	0.0216		0.0246	0.0086

Step-6 & 7-Separation Measures & Closeness Coefficient: In the TOPSIS approach, the positive separation and the negative separation are measured to evaluate the relative closeness coefficients. The results are compiled table-9 by using the mathematical equation number 10 and 11.

Table 9. Separation measures of closeness coefficient

Alternatives	$\theta_{IFS}(\alpha_i, \alpha_i^+)$	$\theta_{IFS}(\alpha_i, \alpha_i^-)$	C_{i^*}	Rank
A_1	0.22334	0.23263	0.51	2
A_2	0.00355	0.44184	0.992	1
A_3	0.23187	0.21879	0.485	3
A_4	0.37167	0.07735	0.172	4

Step-8-Alternative's Rank: The last and final step for the Fuzzy-TOPSIs approach is defining the rank. In the present study, the criteria A_2 is ranked at one; criteria A_1 is ranked at number two; criteria A_3 at number three and finally, the criteria A_4 as listed in table 9.

5. Result and Discussion

Students are the future of a nation. Providing students with the best education possible is very important and can also be considered an investment in the betterment and development of the nation. Though learnings can be deliberated among many followers in multiple ways, i.e., online, hybrid, in-person, etc., Still, classroom education is well recognized. The classroom education not only helps in developing the minds to think out logical solutions but also provides practical exposure, i.e., skill-based learning, to the attendees. Moreover, personal as well as professional skills play an important role in the development of a good career.

For the aforesaid reasons, the role of an institution or university becomes pertinent. Here, it is an exaggeration to say that it seems to be quite complex and difficult to define the role of an institution clearly because it is based on a number of parameters. Even for an institution or university, it becomes prominent to provide not only adequate education but also to fulfil the criteria among all alternatives. The present study reveals the importance of selecting a good university or institution for educational purposes. Most of the young aspirants fail to judge the selection parameter, and thereafter they cheat.

The current study reveals the parameters associated with students' selection of an institution or university prior to making any decision regarding admission enrolment. The ten most important parameters are identified through an extensive literature review and then validated by the experts. Later, the MCDM approach is used to rank all the parameters according to their impact on post-graduation degrees from an institution. The MCDM approach is best suited for problems where the contribution factors in a particular event or activity. Both the academic and non-academic contributors' criteria are considered. In the MCDM approach, the fuzzy TOPSIS method is used to model and rank the linguistic inputs by the experts in the meaningful research. In this study, intuitionistic fuzzy sets are used, which help in providing precise information that further leads to better decision-making. Further, it helps in eliminating or reducing the vagueness of parameters over their effective impact on the decision. Finally, the ranks are defined using Fuzzy-TOPSIS, and the parameter A2, i.e., faculty/research facility, is reported to be the most influencing parameter in the choice or selection of an institution or university. This plays an important role in institutional growth. Further, the concern about the future perspective, i.e., opportunities for career development, also has a specific contribution to the selection of an institution or university. In a nutshell, good selection regarding the institution or university plays a crucial role and, if addressed carefully, will lead to both tangible and intangible benefits for all of its attendees.

Despite the fruitful insights on the topic cited above, the present study has a limitation, i.e., consideration of only 10 parameters. There may be other parameters that can directly or indirectly impact the selection of the university. Other parameters, such as talk delivery, technological advancement, student learning assessments, teaching capabilities, academic ethics confronters, and so on, will be investigated in the future to gain additional insights into selection parameters.

Conclusion

Choosing the right institution that meets the aspirant's needs is the most important factor in obtaining a quality education. The factors influencing the selection of the best educational universities have been considered. Multi-criteria decision-making The TOPSIS method has been used with intuitionistic fuzzy information to rank academic as well as non-academic parameters that help the student get admission to a better university.

References

Alidrisi, H., and Al-Sasi, B. O. (2017). Utilization of energy sources by G20 countries: A TOPSIS-BASED approach. *Energy Sources, Part B: Economics, Planning, and Policy*, 12(11), 964-970.

Atanassov, K. T. (1986). Intuitionistic fuzzy sets. *Fuzzy Sets and Systems*, 20(1), 87-96.

Atanassov, K. T. (1999). Intuitionistic fuzzy sets. *Physical-Verlag Heidelberg* - New York, 1-137.

Atanassov, K., Pasi, G. and Yager, R. R. (2005). Intuitionistic fuzzy interpretations of multi-criteria multi-person and multi-measurement tool decision-making. *International Journal of System Science,* 36(14), 859–868.

Adams, J. and Evenland, V. (2007). Marketing online degree programs: how do traditional residential programs compare? *Journal of Marketing for Higher Education*, 17(1), 67–90.

Broekemier, G. M. and Seshadri, S. (2000). Differences in college choice criteria between deciding students and their parents, *Journal of Marketing for Higher Education*, 9(3), 1–13.

Briggs, S. (2006). An exploratory study of the factors influencing undergraduate student choice: the case of higher education in Scotland. *Studies in Higher Education,* 31(6), 705–722.

Briggs, S. and Wilson, A. (2007). Which university? A study of the influence of cost and information factors on Scottish undergraduate choice. *Journal of Higher Education Policy and Management*, 29(1), 57–72.

Çelikbilek, Y. and Tüysüz, F. (2020). An in-depth review of theory of the TOPSIS method: An experimental analysis. *Journal of Management Analytics*, 7(2), 281-300.

Chakraborty, S. and Chatterjee, P. (2013). Selection of Materials Using Multi-Criteria Decision-Making Methods with Minimum Data. *Decision Science Letters*, 7, 135–48.

Chen, P. (2019). Effects of normalization on the entropy-based TOPSIS method. *Expert Systems with Applications*, 136, 33-41.

Chen T. Y. and Li, C. H. (2010) Determining objective weights with intuitionistic fuzzy entropy measures: A comparative analysis. *Inform Science*, 180(21), 4207–22.

Conard, M. J. and Conard, M. A. (2000). An analysis of academic reputation as perceived by consumers of higher education. *Journal of Marketing for Higher Education*, 9(4), 69–79.

Cremonini, L., Westerheijden, D. and Enders, J. (2008). Disseminating the right information to the right audience: cultural determinants in the use (and misuse) of rankings. *Higher Education*, 55(3), 373–385.

Damary, R.; Markova, T.; Pryadilina, N. (2017). Key challenges of on-line education in multi-cultural context. In: *Social and Behavioral Sciences*, 237, 83-89.

Grzegorzewski, P. (2004) Distances between intuitionistic fuzzy sets and/or interval-valued fuzzy sets based on the Hausd-off metric. *Fuzzy Sets System,* 49(2), 319–28.

Hwang, C. L. and Yoon, K. (1981) Multiple Attribute Decision-making: Methods and Applications. *Springer-Verlag*, New York.

Kun, Y., Zhu, N., Chang, C., Wang, D., Yang, S. and Shengming, M. (2018). A Methodological Concept for Phase Change Material Selection Based on Multi-Criteria Decision-making (MCDM): A Case Study. *Energy*, 165, 1085–96.

Li, X., Wang, K., Liu, L., Xin, J., Yang, H. and Gao, C. (2011). Application of the Entropy Weight and TOPSIS Method in Safety Evaluation of Coal Mines. In: *Procedia Engineering*, 26, 2085–91.

Liu, H. W. and Wang, G. J. (2007) Multi-criteria decision-making methods based on intuitionistic fuzzy sets. *European Journal of Operation Research,* 179, 220–233.

Kumar, R., Kumar, V. and Singh, S. (2016). Relationship establishment between lean manufacturing and supply chain characteristics to study the impact on organizational performance using SEM approach. *International Journal of Value Chain Management*, 7 (4), 352-367.

Kumar, R., Kumar, V. and Singh, S. (2018). Work culture enablers: Hierarchical design for effectiveness and efficiency. *International Journal of Lean Enterprise Research*, 2(3), 189-195.

Kumar, R., Ritu, Sindhwani, R. and Singh, P. L. (2021). Developing the structural model for barriers associated with the CSR using ISM to help in creating the brand image of the manufacturing industry. *International Journal of Advanced Operations Management*, 13(3), 312-330.

Pampaloni, A. M. (2010). The influence of organizational image on college selection: what students seek in institutions of higher education. *Journal of Marketing for Higher Education,* 20(1), 19–48.

Poock, M. C. and Love, P. G. (2001). Factors influencing the program choice of doctoral students in higher education administration. *NASPA Journal*, 38(2), 203–223.

Reese, S. A. (2015). Online learning environments in higher education: Connectivism vs. dissociation. *Education and information technologies*, 20(3), 579-588.

Rouyendegh, B. D., Yildizbasi, A. and Üstünyer, P. (2020). Intuitionistic fuzzy TOPSIS method for green supplier selection problem. *Soft Computing*, 24(3), 2215-2228.

Shapiro, H. B., Lee, C. H., Roth, N. E. W., Li, K., Çetinkaya-Rundel, M., Canelas, D. A. (2017). Understanding the massive open online course (MOOC) student experience: An examination of attitudes, motivations, and barriers. *Computers & Education*, 110, 35-50.

Şengül, Ü., Eren, M., Shiraz, S. E., Gezder, V., and Şengül, A. B. (2015). Fuzzy TOPSIS method for ranking renewable energy supply systems in Turkey. *Renewable Energy*, 75, 617-625.

Shu, M. S., Cheng, C. H. and Chang, J. R. (2006) Using intuitionistic fuzzy sets for fault tree analysis on printed circuit board assembly. *Microelectron Reliability*, 46(12), 2139–2148.

Sindhwani, R., Singh, P.L., Kumar, R., Kumar, B., Mittal, V.K., Dixit, S. & Jindal, A. (2021). Structural Modelling for the Factors of Facility Management in the Healthcare Industry Using the TISM Approach. In: *Multi-Criteria Decision Modelling: Applicational Techniques and Case Studies*, CRC Press, 01-21.

Sindhwani, R., Kumar, R., Behl, A., Singh, P. L., Kumar, A., & Gupta, T. (2021). Modelling enablers of efficiency and sustainability of healthcare: A m-TISM approach. *Benchmarking: An International Journal*, 29(3), 767-92.

Singh, A., Kazi, R. and Divekar, R. (2019). A Conceptual Model on Fee Perception and Quality of Management Education using ISM and Fuzzy MICMAC Analysis. *Prabandhan: Indian Journal of Management*, 12(7), 36-48.

Simoes, C. and Soares, A. M. (2010). Applying to higher education: information sources and choice factors. *Studies in Higher Education*, 35(4), 371–389.

Sojkin, B., Bartkowiak, P. and Skuza, A. (2011). Determinants of higher education choices and student satisfaction: the case of Poland, *Higher Education*, 63(5), 565–581.

Soutar, G. and Turner, J. (2002). Students' preferences for university: a conjoint analysis. *The International Journal of Educational Management*, 16(1), 40–45.

Sun, A. and Chen, X. (2016). Online education and its effective practice: A research review. *Journal of Information Technology Education*, 15.

Szmidt, E. and Kacprzyk, J. (2000). Distances between intuitionistic fuzzy sets. *Fuzzy Sets System*, 114, 505–518.

Szmidt, E. and Kacprzyk, J. (2001). Intuitionistic fuzzy sets in some medical applications. *Lecture Notes in Computer Science,* 2206, 148–151.

Szmidt, E. and Kacprzyk, J. (2004). A similarity measure for intuitionistic fuzzy sets and its application in supporting medical diagnostic reasoning. *Lecture Notes in Computer Science,* 3070, 388–393.

Taylor, J. S., Brites, R., Correia, F., Farhangmehr, M., Ferreira, B. and Machado, M. L. (2008). Strategic enrolment management: improving student satisfaction and success in Portugal. *Higher Education Management and Policy,* 20(1), 129–145.

Tømte, C., Enochsson, A. B., Buskqvist, U. and Kårstein, A. (2015). Educating online student teachers to master professional digital competence: The TPACK-framework goes online. *Computers & Education*, 84, 26-35.

Wang, W.Q. and Xin, X.L. (2005) Distance measure between intuitionistic fuzzy sets. *Pattern Recognition Letter,* 26, 2063–2069.

Warwick, J. and Mansfield, P. M. (2004). Perceived risk in college selection: differences in evaluative criteria used by students and parents. *Journal of Marketing for Higher Education,* 13(1), 101–125.

Xu, Z. S. (2007) Intuitionistic fuzzy aggregation operators. *IEEE Trans Fuzzy Syst*em, 15(6), 1179–1187.

Xu, Z. S. and Yager, R. R. (2006). Some geometric aggregation operators based on intuitionistic fuzzy sets. *International Journal of Gen System*, 35, 417–433.

Xu, Z. S. and Yager, R. R. (2008). Dynamic intuitionistic fuzzy multi-attribute decision-making. *International Journal of Approx Reason*, 48, 246–262.

Xu, Z. and Zhang, X. (2013). Hesitant fuzzy multi-attribute decision-making based on TOPSIS with incomplete weight information. *Knowledge-Based System*, 52, 53–64.

Yazdani, M. and Payam, A. F. (2015). A Comparative Study on Material Selection of Microelectromechanical Systems Electrostatic Actuators Using Ashby, VIKOR and TOPSIS. *Materials & Design,* 65, 328–34.

Zadeh L A. (1965). Fuzzy sets. *Information Control*, 8(3), 338–56.

Zadeh, L. A. (1973). Outline of a new approach to the analysis of complex systems and decision processes. *IEEE Transactions on Systems, Man, and Cybernetics, SMC, 3*(1), 28-44.

Zhang, H., Chao-lin, G., Lu-wen, G. and Zhang, Y. (2011). The Evaluation of Tourism Destination Competitiveness by TOPSIS & Information Entropy – A Case in the Yangtze River Delta of China. *Tourism Management*, 32 (2), 443–51.

Zeleny, M. (1976). The attribute-dynamic attribute model. *Management Science*, 23, 12–26.

Chapter 5

Generalized Information Theoretic Measures with Applications in Decision-Making

Vijay Kumar*
and Sarika Jain
Manav Rachna International Institute of Research and Studies,
Faridabad, Haryana, India

Abstract

Entropy is a tool by which uncertainty in any system has been measured. In 1948, C. E. Shannon introduced the concept of entropy in probability distribution, which gave birth to a separate discipline known as Information Theory. After that, many researchers proposed generalized entropies that contributes in the development of the theory. In 1965, L A Zadeh introduced the concept of fuzzy set and soon after its incorporation, he proposed the entropy of a fuzzy event in 1969. After that many generalizations of fuzzy sets have been introduced, which have applications in decision-making problems. Both these disciplines give birth to a new discipline known as fuzzy information theory. One such generalization is Intuitionistic fuzzy (IF) set, introduced by Atanassov in 1986. In this chapter, a new generalized trigonometric information theoretic measure has been proposed, known as trigonometric intuitionistic fuzzy entropy measure. The validation of the proposed measure has been discussed with the help of mathematical proofs of the theorems. Finally, the consistency of the measure has been discussed by taking numerical examples. The proposed measure has been used for decision-making problems across disciplines.

---

* Corresponding Author's Email: drvijaykumarsudan@gmail.com.

In: Understanding Information Entropy
Editor: Vijay Kumar
ISBN: 979-8-88697-839-1

Keywords: intuitionistic fuzzy sets (IFSs), entropy measure, trigonometric intuitionistic fuzzy entropy, fuzzy information measure

1. Introduction

In real world situations, we deal with deterministic and non-deterministic phenomena. Out of this, non-deterministic phenomena face more challenges in the study of decision-making processes. The study of non-deterministic phenomena has been taken up by statistics, if the occurrences of patterns are stable. However, in reality, there are a large number of imprecise and vague notations, which are not well defined and beyond the scope of precise notations of mathematics. For such vague situations, (Zadeh 1965) introduced Fuzzy Set Theory, as a classical set theory generalization for describing imprecise and vague situations. Fuzzy set theory has the flexibility to represent incomplete information about a situation. It is a class that considers partial membership as real numbers in [0,1]. From the past more than five decades, researchers are engaged in producing new developments in the theory of fuzzy sets. This concept has applications ranging from pure mathematical subjects such as analysis, algebra, logic, calculus etc to interdisciplinary subjects such as decision theory, graph theory, operations research, information theory, coding theory, pattern recognition, cluster analysis etc. Consequently, fuzzy set theory and its generalizations has been emerged as potential area of interdisciplinary research among the researchers. Intuitionistic fuzzy set (IFS) theory (Atanassov 1986) is the generalization of fuzzy set theory, characterized by degree of membership and non-membership functions respectively. It is known that IFSs describe the fuzzy character more comprehensively in some special situations and therefore, are a more useful tool to deal with fuzzy information. In this changing world, technology is transforming everything and many decision support tools have been developed, which make the life of decision makers easy and helped them to take decisions under favorable or unfavorable circumstances. With the development of this, computational techniques such as fuzzy theory and its generalization provides hand held support to the decision makers that allow the discipline to grow exponentially across disciplines. Fuzzy set and its generalizations have applications in many domains such as medical diagnosis, decision theory, pattern recognition etc. Among this, Intuitionistic fuzzy set (IFS) describes the fuzzy character more comprehensively in special situations and has variety of applications in decision theory ranging from market prediction to medical diagnosis. In this work, medical decision-making support system for various life-threatening diseases has been considered, which supports the doctors and the patients in diagnostic process. The complex system of human body may not be handled through traditional approaches as such systems have certain amount of fuzziness, therefore, the precise mathematical notations are not the eligible candidate to deal with large number of variables in decision-making process.

2. Information Theoretic Measures

We would start with probabilistic approach and then take up with fuzzy and then intuitionistic fuzzy approaches.

2.1. Probability Approach

For a random variable, entropy is a quantitative measure, used to calculate uncertainty. Shannon (1948), coined the term entropy, which gave birth to a separate discipline namely Information theory (Ash 1965). He defined the concepts of information in mathematical insight and proved many general results and their consequences. Since then, entropy has been extensively used in different domains including quantum information theory, communication theory, image recognition, finance, decision-making etc. Shannon entropy has generalizations proposed by (Rényi 1961), (Tsallis 1988), (Havrda & Charvát 1967), (Pal & Pal 1989) etc. These entropies have applications in various domains of study such as pattern recognition, image extraction, feature selection, decision-making etc.

2.2. Fuzzy Approach

Zadeh (1969) anticipated that fuzzy theory can handle the problems of medical diagnosis very well to deal with imprecise and uncertain information gathered by humans. The primary characteristic of fuzzy theory is that it accepts the information in both qualitative and quantitative ways and can develops an ecosystem, which describe it by simple human-friendly rules. In any diagnostic process, generally medical knowledge base is imprecise and uncertain, which is a relationship between the symptoms and the disease. Information about patients is gathered from past history, clinical examination, pathological tests results and other available investigative procedures. This information has some degree of incompleteness and certain things are overlooked or wrongly interpreted during investigation, resulting wrong diagnosis and the execution of wrong treatment of the actual disease. In such situations, many fuzzy based models intervene and can help in establishing certain relationships such as, variation of important symptoms, pattern of symptoms of different diseases, relationship between the diseases and their stages, initial and final diagnosis. These models form the initial base for information about the diseases, which may be iterated and configured as per the requirement. Sanchez (1977a, 1977b) proposed the fuzzy based model, which represented the knowledge base by establishing a fuzzy max-min relationship between the symptoms and the disease.

2.3. Intuitionistic Fuzzy Approach

Many scholars have conducted researches on entropy measures based on IFS. Burillo & Bustince (1996) introduced the first axiom of IF entropy. It was a distance measure, which measures the degree of intuitionism associated with IFS. Szmidt & Kacprzyk (2001a) extended the definition of IF entropy, which was the extension of (De Luca & Termini 1972) and proposed a new entropy measure for IFS. Hung & Yang (2004) proposed IF entropy measure for two families of entropies and defines their axiomatic definition and properties. Another measure of IF entropy proposed by (Vlachos & Sergiadis 2007), which revealed the connection between the intuitive and the mathematics of entropy. Zhang & Jiang (2008) defined a measure of IF entropy by generalizing (De Luca & Termini 1972) logarithmic fuzzy entropy. Many researchers have defined the IF entropy from different viewpoints, such as probabilistic viewpoint, non-probabilistic viewpoint and geometric viewpoint. From the past few years, many researchers have proposed different entropy measures, including (Verma & Sharma 2013), (Mishra 2016), (Wei 2012), (Wang & Zhang 2009a, 2009b), (Kumar, Arora & Pal 2016), (Kumar & Jain 2018), (Liu & Ren 2014). They have applied these entropies in different real world problems.

3. Decision-Making

In decision-making, selection of best alternative from a finite number of available alternatives have been taken place. It is an art and a common activity, which is used in every sphere of human functionality. Decision-making has applications in business, finance, management, economics, engineering, medical science etc. Researchers across various disciplines have shown great interest in solving problems related to decision-making. In real life situations, usually ambiguous or imprecise information is available and in such cases, fuzzy and IFS theory have emerged as powerful tools. To solve such real life situations flavored with information measures in fuzzy and IFS environment, mark the beginning of new disciplines fuzzy information theory and IF information theory.

In order to define the entropy function, fuzzy set and IFS theoretic approaches are found useful in many real life situations. For this, we must have to suggest a correct mathematical model, which acts as tool to solve real life problems. Many decision-making techniques have been developed to handle real life situations.

Decision-making is a vital activity to make decisions based on the judgment and is a cognitive process. In decision-making, one alternative is selected among the available alternatives. In deterministic process, decision-making can be performed by means of an optimization problem but in non-deterministic problems, making decisions under uncertainty is a challenge. Under such circumstances, decision makers face a lot of confusion and ambiguity in evaluating the criteria of alternatives. In such situation, fuzzy set theory can model and handle the vague information and situation very well.

There are varieties of situations in which decision-making can be performed such as: Individual decision-making, Multi-person decision-making, Multi-stage decision-making and Multi-criteria decision-making (MCDM). Out of these, MCDM is used frequently in real life situations and can handle the problems of incompleteness.

4. Multi Criteria Decision-Making (MCDM)

Multiple criteria decision-making (MCDM) is a branch of decision-making to design computational tools, which supports the evaluation of performance criteria with a aim to solve real world issues across the domains related to the areas including health, energy, supply chain management, quality management, manufacturing systems, etc. During past six decades, MCDM techniques have grown many folds, which provide platform to classify alternatives and rank them according to the preference. Traditional decision-making problem has been structured into three phases as given in Figure 1.

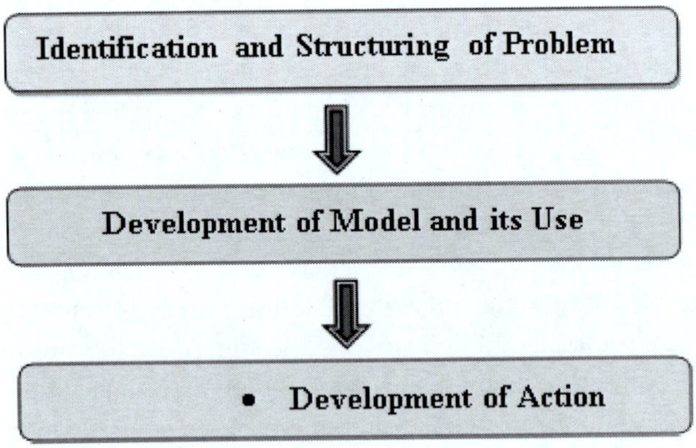

Figure 1. Phases of MCDM Problems.

MCDM techniques have been used by the researchers in various domains of the universe, but the prominent MCDM techniques are laced with fuzzy theory. Atanassov (1986) proposed the concept of IFS, a generalization of fuzzy set, which is characterized by membership and non-membership functions. Later, (Atanassov 1999) and (Bustince & Burilo 1996) developed the theory to handle more complex decision-making problems. Gau & Buehrer (1993) introduced the concept of vague set, which is an equivalence of IFS by (Bustine & Burillo 1996). De, Biswas & Roy (2001, 2003) developed new categories of IFS operators such as concentration, dilation and normalization. Based on vague sets and IFS, (Chen & Tan 1994) and (Szmidt & Kacprzyk 2001c) developed some techniques for handling MCDM problems under fuzzy environment. Xu & Yager (2006) developed some aggregation operators, which include IFWA operator, IFWG operator, IFOWG operator and IFHG operator for handling MCDM problems under IFS environment. Liu & Wang (2007) developed evaluation function, commonly known as score functions, which measure the degree of satisfaction among the alternatives. Uncertain information cannot be estimated with exact numerical values but with linguistic terms interpreted by natural languages.

MCDM problem establishes the connection among the set of alternatives with respect to given evaluating factors. Experts have been given the task to select best alternative among the set of available ones and evaluate the alternatives on the basis of some criteria with an aim to provide better solution. To produce overall ranking of the preferences and to achieve tangible goals, MCDM techniques have been put forth. Moreover, MCDM methods identify the set of criteria by which alternatives are to be compared.

The medical information based computational techniques optimize decisions in each step to bridge up the gaps which occur during the treatment. These techniques take care of all types of complexities and navigate the patients and doctors by offering other candidate alternatives. Any form of illness present in human body is deadly and the trajectory of the disease from diagnosis to the treatment passes through certain challenging decisions. Information based management system could be the key factor in understanding the diagnosis of diseases, making decisions and if possible, reconfigure the treatment. Many decision-making models focus on one criterion and have limited applications; therefore, they fail in most of the real-life situations.

5. Some Preliminaries

5.1. Crisp Set

Crisp set is a conventional bivalent set. For a crisp set A of X,

$\exists \ f_A(x): X \rightarrow \{0,1\}$; where, X is universal set.

and $$f_A(x) = \begin{Bmatrix} 1, & if\ x \in A \\ 0, & if\ x \notin A \end{Bmatrix}$$

In this set, element either is a member of set or not.

5.2. Fuzzy Set (FS)

Zadeh (1965), FS is an extension of the classical set, whose elements have degree of membership.

A fuzzy set A is defined in a discrete universe of discourse $X=\{x_1,x_2,...,x_n\}$ as: $A=\{<x, \mu_A(x)> : x \in X\}$ where,

$\mu_A : X \rightarrow [0,1]$ is the membership function of set A and

$\mu_A(x)$ is called the grade of membership of $x \in X$ in A

Fuzzy sets have many generalizations. In this work, intuitionistic fuzzy sets (IFSs), have been used for the purpose of decision-making.

5.3. Intuitionistic Fuzzy Sets (IFSs) and Its Operations

Atanassov (1986) An IFS A in $X=\{x_1,x_2,...,x_n\}$ is given as:

$A=\{<x, \mu_A(x), \nu_A(x)> | x \in X\}$; where

$\mu_A : X \rightarrow [0,1]$ and $\nu_A : X \rightarrow [0,1]$ are the membership and non-membership functions of set A.

$\pi_A(x)=1-\mu_A(x)-\nu_A(x)$ is intuitionistic index or hesitation index of x in A.

In limiting case, if $\pi_A(x)=0$, IFS reduces automatically to FS.

Let $A, B \in X$ be two IFS s.t.

$$A=\{<x, \mu_A(x), \nu_A(x)> | x \in X\},$$
$$B=\{<x, \mu_B(x), \nu_B(x)> | x \in X\}$$

Atanassov (1999) and (Szmidt & Kacprzyk 2001a), operations on IFS are given as:

a. $A \subseteq B$ *iff* $\mu_A(x) \leq \mu_B(x)$ and $\nu_A(x) \geq \nu_B(x)$ $\forall x \in X$;

b. $A = B$ *iff* $A \subseteq B$ and $B \subseteq A$;

c. $A^C = \{<x, \nu_A(x), \mu_A(x) > | x \in X\}$;

d. $A \cup B = \{<x, (\mu_A(x) \vee \mu_B(x)), (\nu_A(x) \wedge \nu_B(x)) > | x \in X\}$;

e. $A \cap B = \{<x, (\mu_A(x) \wedge \mu_B(x)), (\nu_A(x) \vee \nu_B(x)) > | x \in X\}$;

f. $\urcorner A = \{<x, \mu_A(x), 1 - \mu_A(x) > | x \in X\}$;

g. $\Diamond A = \{<x, 1 - \nu_A(x), \nu_A(x) > | x \in X\}$;

h. $A @ B = \left\{<x, \frac{\mu_A(x) + \mu_B(x)}{2}, \frac{\nu_A(x) + \nu_B(x)}{2} > | x \in X\right\}$;

i. $A \leq B$ *iff* $\mu_A(x) \leq \mu_B(x)$ and $\nu_A(x) \geq \nu_B(x)$ for $\mu_B(x) \leq \nu_B(x)$

or $\mu_A(x) \geq \mu_B(x)$ and $\nu_A(x) \leq \nu_B(x)$ for $\mu_B(x) \geq \nu_B(x)$ $\forall x \in X$.

5.4. Intuitionistic Fuzzy Number (IFN) and Its Operations

Dubois & Prade (1980) introduced FN and (Burillo, Bustince & Mohedano 1994) introduced IFN. IFN $\tilde{A}$ is an IFS satisfying the following conditions as:

5.4.1: IF subset of the real line.

5.4.2: Normal i.e., for any $x_0 \in R$ such that $\mu_{\tilde{A}}(x_0) = 1$ so $\nu_{\tilde{A}}(x_0) = 0$

5.4.3: A convex set for $\mu_{\tilde{A}}(x)$

i.e., $\mu_{\tilde{A}}(\lambda x_1 + (1-\lambda)x_2) \geq \min(\mu_{\tilde{A}}(x_1) \leq, \mu_{\tilde{A}}(x_2))$; $\forall x_1, x_2 \in R$, $\lambda \in [0,1]$

5.4.4: A concave set for $\nu_{\tilde{A}}(x)$

i.e., $\nu_{\tilde{A}}(\lambda x_1 + (1-\lambda)x_2) \leq \max(\nu_{\tilde{A}}(x_1), \nu_{\tilde{A}}(x_2))$; $\forall x_1, x_2 \in R$, $\lambda \in [0,1]$

Let $A = (\mu_A, \nu_A)$ be an IFN with $0 \leq \mu_A(x) + \mathrm{v}_A(x) \leq 1$; $\mu_A, \mathrm{v}_A \in [0, 1]$ for $x \in X$

Yazdani et al., (2017) the operations of IFNs $A_1 = (\mu_{A_1}, \nu_{A_1})$, $A_2 = (\mu_{A_2}, \nu_{A_2})$ and $A_3 = (\mu_{A_3}, \nu_{A_3})$ are defined as:

a. $A_1 + A_2 = (\mu_{A_1} + \mu_{A_2} - \mu_{A_1}\mu_{A_2}, \nu_{A_1}\nu_{A_2})$

b. $A_1 \times A_2 = (\mu_{A_1}\mu_{A_2}, \nu_{A_1} + \nu_{A_2} - \nu_{A_1}\nu_{A_2})$

c. $\lambda A = (1-(1-\mu_A)^{\lambda}, \nu_A^{\lambda}); \lambda > 0$

d. $A^{\lambda} = (\mu_A^{\lambda}, 1-(1-\nu_A)^{\lambda}); \lambda > 0$

6. Intuitionistic Fuzzy Entropy (IFE) Measure

Shannon (1948) introduced a new discipline as information theory and after that entropy is used in almost all the domains of the universe. Based on IFSs theory, our work is in information theoretic measures like entropy and cross entropy with applications in decision-making processes. For a random variable, entropy is a quantitative measure used to calculate uncertainty. Shannon (1948), coined the term entropy, which gave birth to many other disciplines and attracts researchers from different domains of study such as pattern recognition, image extraction, feature selection, decision-making etc. Many scholars are conducting researches in entropy measures on IFS. The first axiom of IF entropy measures the intuitionistic degree associated with IFS and is known as distance measure. A new entropy measure for IFS has been proposed by (Szmidt & Kacprzyk 2001a, 2001b, 2001c), which is the extension of (De Luca & Termini 1972). Zadeh (1965) defined fuzzy sets for non-statistical modeling and incomplete information data (Yazdani et al., 2017). Since its inception, fuzzy sets became an interesting area of research and applied for decision-making purposes in every sphere of the universe. These sets deal with the situations in which complete information is not known (Fan & Ma 2002; Hooda & Mishra 2015; Das, Guha & Dutta 2016). Many researchers have applied fuzzy sets in their research and get remarkable results, which give more recognition to the theory. There are many generalizations of fuzzy sets. Zadeh (1968) proposed fuzzy entropy based on the probabilistic framework as a measure of fuzziness of a fuzzy set. Many researchers applied fuzzy entropy in statistical inference (Zamanzade & Arghami 2011, 2012; Zamanzade, 2014; Zamanzade & Mahdizadeh 2016, 2017). The concept of entropy has been given by Shannon in 1948, which is statistical in nature. De Luca & Termini (1972) defined fuzzy entropy, which was based on Shannon measure of entropy and provided the axioms used to define new fuzzy entropy measures. Many researchers have employed these axioms to propose new fuzzy entropy. Researchers such as (Parkash, Sharma & Mahajan 2008), (Verma & Sharma 2013), (Hooda & Mishra 2015) proposed several fuzzy based trigonometric entropies in terms of sine and cosine functions and studied their properties as well. They extended Shannon measure of entropy to measure fuzziness and characterized it by two ways, one way was to introduce fuzzy entropy as a measure of the degree of fuzziness and the other was that they defined the properties of fuzzy entropy. Atanassov (1986) introduced the generalized version of fuzzy set known as IFS by including membership function, non-membership function and hesitation index or intuitionistic index. From the past many decades, researchers have proposed entropies based on IFSs, which have applications in the fields of decision-making, optimization, pattern recognition etc. To establish a mathematical relationship between the entropy for FS and IFS, many new measures of IF entropy has been proposed by many researchers such as, (Burillo & Bustince 1996), (De Luca & Termini 1972), (Vlachos & Sergiadis 2007), (Wei 2012), (Pal & Pal 1989), (Verma & Sharma 2013), (Ye 2010), (Liu & Ren 2014), (Mishra 2016). These entropies satisfy all the axioms for the IF entropy and can be applied in many areas of research.

In this chapter, new trigonometric entropy on IFS has been proposed that validates all the axioms given by (Szmidt & Kacprzyk 2001a) along with other related mathematical properties.

6.1. Axioms for Candidate Entropy Measure

According to (Szmidt & Kacprzyk 2001a), a function should satisfy the following axioms to be a valid measure of IF Entropy:

6.1.1: $E(A)=0$ *iff* $\mu_A(x_i)=0, \nu_A(x_i)=1$ or $\mu_A(x_i)=1, \nu_A(x_i)=0$ $\forall\, x_i \in X$;

i.e., A is a crisp set.

6.1.2: $E(A)=1$ *iff* $\mu_A(x_i)=\nu_A(x_i)$ $\forall\, x_i \in X$;

6.1.3: $E(A)=E(A^C)$

6.1.4: $E(A)\le E(B)$ iff $A \le B$

In the next section, trigonometric based IF entropy measure has been proposed.

6.2. Proposed Trigonometric Intuitionistic Fuzzy Entropy Measure

Let $A = \{< x, \mu_A(x), \nu_A(x) > | \, x \in X\}$ be an IFS, where $X = \{x_1, x_2, ..., x_n\}$,

For the IFS A, the trigonometric IFE measure is defined as:

$$E(A) = \frac{1}{n}\sum_{i=1}^{n}\left[\cos\frac{\pi}{2}\left(\left|\mu_A(x_i) - \nu_A(x_i)\right|(1 - \pi_A(x_i))\right)\right]$$

Or, $E(A) = \frac{1}{n}\sum_{i=1}^{n}\left[\cos\frac{\pi}{2}\left(\left|\mu_A{}^2(x_i) - \nu_A{}^2(x_i)\right|\right)\right]$ as $\pi_A(x) = 1 - \mu_A(x) - \nu_A(x)$

In the next section, the proof of the theorems (6.3-6.4) have been discussed to validate proposed entropy.

6.3. Theorem

The trigonometric IF entropy measure satisfies axioms given by (Szmidt & Kacprzyk 2001a).

Proof

For IFS $A = \{< x, \mu_A(x), \nu_A(x) > | \, x \in X\}$

Since $\mu_A(x_i), \nu_A(x_i) \in [0,1] \; \forall i$ we have, $0 \le \left|\mu_A{}^2(x_i) - \nu_A{}^2(x_i)\right| \le 1$

Also, $0 \le \cos\frac{\pi}{2}\left(\left|\mu_A{}^2(x_i) - \nu_A{}^2(x_i)\right|\right) \le 1$

$$\Rightarrow \; 0 \le \frac{1}{n}\sum_{i=1}^{n}\left[\cos\frac{\pi}{2}\left(\left|\mu_A{}^2(x_i) - \nu_A{}^2(x_i)\right|\right)\right] \le 1$$

Thus, $0 \leq E(A) \leq 1$

To justify the validity claim of the proposed entropy measure, the axioms proposed by (Szmidt & Kacprzyk 2001a) need to be satisfied.

The detailed proof of each axiom has been given as:

Axiom 6.3.1

$$E(A) = 0 \textit{ iff } \mu_A(x_i) = 0, \nu_A(x_i) = 1 \text{ or } \mu_A(x_i) = 1, \nu_A(x_i) = 0 ; \forall x_i \in X$$

Proof

Let A be a crisp set, $\forall x_i \in X$, We have:

$$\mu_A(x_i) = 0, \nu_A(x_i) = 1 \text{ or } \mu_A(x_i) = 1, \nu_A(x_i) = 0$$

$$\Rightarrow E(A) = \frac{1}{n}\sum_{i=1}^{n}\left[\cos\frac{\pi}{2}\right] \Rightarrow E(A) = 0$$

Conversely, If $E(A) = 0$

$$\Rightarrow \frac{1}{n}\sum_{i=1}^{n}\left[\cos\frac{\pi}{2}\left(\left|\mu_A^2(x_i) - \nu_A^2(x_i)\right|\right)\right] = 0$$

$$\Rightarrow \cos\frac{\pi}{2}\left(\left|\mu_A^2(x_i) - \nu_A^2(x_i)\right|\right) = 0 \;\forall x_i \in X,$$

$$\Rightarrow \left|\mu_A^2(x_i) - \nu_A^2(x_i)\right| = 1 \;\forall x_i \in X,$$

Thus either $\mu_A(x_i) = 0, \nu_A(x_i) = 1$ or $\mu_A(x_i) = 1, \nu_A(x_i) = 0$

Thus, A be a crisp set.

Axiom 6.3.2

$$E(A)=1 \quad iff \quad \mu_{\mathrm{A}}(x_i)=\nu_{\mathrm{A}}(x_i); \quad \forall\, x_i \in X$$

Proof

Let $\mu_{\mathrm{A}}(x_i)=\nu_{\mathrm{A}}(x_i),\ \forall\ x_i \in X$

$$\Rightarrow \left|\mu_A^{\,2}(x_i)-\nu_A^{\,2}(x_i)\right|=0 \ \Rightarrow \cos\frac{\pi}{2}\left(\left|\mu_A^{\,2}(x_i)-\nu_A^{\,2}(x_i)\right|\right)=1$$

$$\Rightarrow \frac{1}{n}\sum_{i=1}^{n}\left[\cos\frac{\pi}{2}\left(\left|\mu_A^{\,2}(x_i)-\nu_A^{\,2}(x_i)\right|\right)\right]=1 \ \Rightarrow E(A)=1$$

Conversely

Given, $E(A)=1 \ \Rightarrow \ \frac{1}{n}\sum_{i=1}^{n}\left[\cos\frac{\pi}{2}\left(\left|\mu_A^{\,2}(x_i)-\nu_A^{\,2}(x_i)\right|\right)\right]=1$

$$\Rightarrow \sum_{i=1}^{n}\left[\cos\frac{\pi}{2}\left(\left|\mu_A^{\,2}(x_i)-\nu_A^{\,2}(x_i)\right|\right)\right]=n \ \Rightarrow \ \cos\frac{\pi}{2}\left(\left|\mu_A^{\,2}(x_i)-\nu_A^{\,2}(x_i)\right|\right)=1$$

$$\Rightarrow \left|\mu_A^{\,2}(x_i)-\nu_A^{\,2}(x_i)\right|=0 \ \Rightarrow \ \mu_{\mathrm{A}}(x_i)=\nu_{\mathrm{A}}(x_i),\ \forall\ x_i \in X$$

Axiom 6.3.3 $E(A)=E(A^C)$

Proof

Given that $A^C=\{<x, \nu_A(x), \mu_A(x)>|\, x\in X\}$

We have $E(A^C) = \frac{1}{n}\sum_{i=1}^{n}\left[\cos\frac{\pi}{2}\left(\left|\nu_A^2(x_i) - \mu_A^2(x_i)\right|\right)\right]$

$$= \frac{1}{n}\sum_{i=1}^{n}\left[\cos\frac{\pi}{2}\left(\left|\mu_A^2(x_i) - \nu_A^2(x_i)\right|\right)\right] = E(A)$$

$$\Rightarrow E(A) = E(A^C)$$

Axiom 6.3.4 $E(A) \leq E(B)$ *iff* $A \leq B$

Proof

Let us construct a function

$$f(x, y) = \cos\frac{\pi}{2}\left(x^2 - y^2\right) \text{ where } x, y \in [0, 1]$$

To prove (axiom 6.4.4), for $x \leq y$, it is sufficient to prove that $f(x, y)$ is increasing with respect to x and decreasing with respect to y respectively.

Thus, $\frac{\partial f}{\partial x} = \pi\, x \sin\frac{\pi}{2}\left(y^2 - x^2\right)$ and $\frac{\partial f}{\partial y} = \pi\, y \sin\frac{\pi}{2}\left(x^2 - y^2\right)$

To find the optimal points,

Taking $\frac{\partial f}{\partial x} = 0$ and $\frac{\partial f}{\partial y} = 0$,

$$\frac{\partial f}{\partial x} = 0 \Rightarrow \pi\, x \sin\frac{\pi}{2}\left(y^2 - x^2\right) = 0 \Rightarrow \sin\frac{\pi}{2}\left(y^2 - x^2\right) = 0$$

$$\Rightarrow y^2 - x^2 = 0 \Rightarrow x = y$$

Also, $\frac{\partial f}{\partial y} = 0 \Rightarrow \pi\, y \sin\frac{\pi}{2}\left(x^2 - y^2\right) = 0 \Rightarrow \sin\frac{\pi}{2}\left(x^2 - y^2\right) = 0$

$\Rightarrow x^2 - y^2 = 0 \;\Rightarrow\; x = y$

Hence, the critical point is $x = y$

when $x \le y$, $\frac{\partial f}{\partial x} \ge 0$ and $\frac{\partial f}{\partial y} \le 0$

Thus, $\forall\; x, y \in [0, 1]$ and for $x \le y$,

$f(x, y)$ is increasing and decreasing with respect to x and y respectively.

Using this concept on the proposed entropy,

when $\mu_{\mathrm{B}}(x_i) \le \nu_{\mathrm{B}}(x_i)$

$\mu_{\mathrm{A}}(x_i) \le \mu_{\mathrm{B}}(x_i)$ and $\nu_{\mathrm{A}}(x_i) \ge \nu_{\mathrm{B}}(x_i)$ is satisfied

$$\Rightarrow f(\mu_{\mathrm{A}}(x_i),\ \nu_{\mathrm{A}}(x_i)) \le f(\mu_{\mathrm{B}}(x_i),\ \nu_{\mathrm{B}}(x_i))$$

$$\Rightarrow \cos\frac{\pi}{2}\left(\left|\mu_A^{\,2}(x_i) - \nu_A^{\,2}(x_i)\right|\right) \le \cos\frac{\pi}{2}\left(\left|\mu_B^{\,2}(x_i) - \nu_B^{\,2}(x_i)\right|\right)$$

$$\Rightarrow \frac{1}{n}\sum_{i=1}^{n}\left[\cos\frac{\pi}{2}\left(\left|\mu_A^{\,2}(x_i) - \nu_A^{\,2}(x_i)\right|\right)\right] \le \frac{1}{n}\sum_{i=1}^{n}\left[\cos\frac{\pi}{2}\left(\left|\mu_B^{\,2}(x_i) - \nu_B^{\,2}(x_i)\right|\right)\right]$$

$$\Rightarrow E(A) \le E(B)$$

Similarly, for $x \ge y$, $\frac{\partial f}{\partial x} \le 0$ and $\frac{\partial f}{\partial y} \ge 0$

Thus, $\forall\, x, y \in [0,1]$ and for $x \geq y$, $f(x,y)$ is decreasing and increasing with respect to x and y respectively.

To validate this concept on the proposed entropy, when $\mu_{\mathrm{B}}(x_i) \geq \nu_{\mathrm{B}}(x_i)$

$\mu_{\mathrm{A}}(x_i) \geq \mu_{\mathrm{B}}(x_i)$ and $\nu_{\mathrm{A}}(x_i) \leq \nu_{\mathrm{B}}(x_i)$ is satisfied

$$\Rightarrow f\left(\mu_{\mathrm{A}}(x_i),\ \nu_{\mathrm{A}}(x_i)\right) \leq f\left(\mu_{\mathrm{B}}(x_i),\ \nu_{\mathrm{B}}(x_i)\right)$$

$$\Rightarrow \cos\frac{\pi}{2}\left(\left|\mu_A^{\,2}(x_i) - \nu_A^{\,2}(x_i)\right|\right) \leq \cos\frac{\pi}{2}\left(\left|\mu_B^{\,2}(x_i) - \nu_B^{\,2}(x_i)\right|\right)$$

$$\Rightarrow \frac{1}{n}\sum_{i=1}^{n}\left[\cos\frac{\pi}{2}\left(\left|\mu_A^{\,2}(x_i) - \nu_A^{\,2}(x_i)\right|\right)\right] \leq \frac{1}{n}\sum_{i=1}^{n}\left[\cos\frac{\pi}{2}\left(\left|\mu_B^{\,2}(x_i) - \nu_B^{\,2}(x_i)\right|\right)\right]$$

$$\Rightarrow E(A) \leq E(B)$$

Hence, if $A \leq B$, then $E(A) \leq E(B)$

Conversely, If $E(A) \leq E(B)$

$$\Rightarrow \frac{1}{n}\sum_{i=1}^{n}\left[\cos\frac{\pi}{2}\left(\left|\mu_A^{\,2}(x_i) - \nu_A^{\,2}(x_i)\right|\right)\right] \leq \frac{1}{n}\sum_{i=1}^{n}\left[\cos\frac{\pi}{2}\left(\left|\mu_B^{\,2}(x_i) - \nu_B^{\,2}(x_i)\right|\right)\right]$$

$$\Rightarrow \cos\frac{\pi}{2}\left(\left|\mu_A^{\,2}(x_i) - \nu_A^{\,2}(x_i)\right|\right) \leq \cos\frac{\pi}{2}\left(\left|\mu_B^{\,2}(x_i) - \nu_B^{\,2}(x_i)\right|\right),\ \forall\ x_i \in X$$

$$\Rightarrow \left|\mu_A^{\,2}(x_i) - \nu_A^{\,2}(x_i)\right| \geq \left|\mu_B^{\,2}(x_i) - \nu_B^{\,2}(x_i)\right|,\ \forall\ x_i \in X$$

Then either $\mu_{\mathrm{A}}(x_i) \geq \mu_{\mathrm{B}}(x_i)$ and $\nu_{\mathrm{A}}(x_i) \leq \nu_{\mathrm{B}}(x_i)$ is satisfied

or $\mu_{\mathrm{A}}(x_i) \leq \mu_{\mathrm{B}}(x_i)$ and $\nu_{\mathrm{A}}(x_i) \geq \nu_{\mathrm{B}}(x_i)$ is satisfied then $A \leq B$

Thus, $E(A) \leq E(B)$ *iff* $A \leq B$

From axioms (6.4.1-6.4.4), it has been concluded that the proposed entropy is the valid entropy, as it satisfies the axioms given by (Szmidt & Kacprzyk 2001a).

Special case: If $\pi_A(x_i) = 0 \,; \forall\, x_i \in X$,

Since $\pi_A(x) = 1 - \mu_A(x) - \nu_A(x) \;\; \Rightarrow 1 - \mu_A(x) - \nu_A(x) = 0$

$$\Rightarrow \mu_A(x) + \nu_A(x) = 1 \;\Rightarrow \nu_A(x) = 1 - \mu_A(x)$$

$$\Rightarrow \left|\mu_A^{\,2}(x_i) - \nu_A^{\,2}(x_i)\right| = \left|\mu_A^{\,2}(x_i) - (1 - \mu_A(x_i))^2\right| \; = \left|2\mu_A(x_i) - 1\right|$$

Thus $E(A) = \frac{1}{n}\sum_{i=1}^{n}\left[\cos\frac{\pi}{2}\left(\left|\mu_A^{\,2}(x_i) - \nu_A^{\,2}(x_i)\right|\right)\right] = \frac{1}{n}\sum_{i=1}^{n}\left[\cos\frac{\pi}{2}\left(\left|2\mu_A(x_i) - 1\right|\right)\right]$

which is a fuzzy entropy (FE) measure.

Thus, For $\pi_A(x_i) = 0 \;\; IFE \rightarrow FE$

In the next theorem (6.5.), a property for the existence of the proposed trigonometric IF entropy measure has been proved.

6.4. Theorem

Let A and B be two IFSs over the universal set $X = \{x_1, x_2, \ldots, x_n\}$, s.t.

$A(x_i) = \{< x_i, \mu_A(x_i), \nu_A(x_i) > | \, x_i \in X\}$ and

$B(x_i) = \{< x_i, \mu_B(x_i), \nu_B(x_i) > | \, x_i \in X\}$,

whenever $\forall\, x_i \in X$, either $A \subseteq B$ or $A \supseteq B$ are satisfied,

then $E(A \cup B) + E(A \cap B) = E(A) + E(B)$

Proof

Let us divide X into two parts as X_1 and X_2, such that

$$X_1 = \{x_i \in X : A(x_i) \subseteq B(x_i)\} \text{ and } X_2 = \{x_i \in X : A(x_i) \supseteq B(x_i)\}$$

s.t. $\mu_A(x_i) \le \mu_B(x_i)$ and $\nu_A(x_i) \ge \nu_B(x_i);\ \forall\ x_i \in X_1$

also $\mu_A(x_i) \ge \mu_B(x_i)$ and $\nu_A(x_i) \le \nu_B(x_i)$; $\forall\ x_i \in X_2$

Since $A \cup B = \{< x, (\mu_A(x) \vee \mu_B(x)), (\nu_A(x) \wedge \nu_B(x)) > | \, x \in X\}$

Thus $A \cup B = \{< x, \mu_B(x), \nu_B(x) > | \, x \in X_1\}$

and $A \cup B = \{< x, \mu_A(x), \nu_A(x) > | \, x \in X_2\}$

Since, $E(A \cup B) = \frac{1}{n}\sum_{i=1}^{n}\left[\cos\frac{\pi}{2}\left(\left|\mu_{A\cup B}{}^2(x_i) - \nu_{A\cup B}{}^2(x_i)\right|\right)\right]$

$$= \frac{1}{n}\left[\sum_{x_i \in X_1}\cos\frac{\pi}{2}\left(\left|\mu_B{}^2(x_i) - \nu_B{}^2(x_i)\right|\right) + \sum_{x_i \in X_2}\cos\frac{\pi}{2}\left(\left|\mu_A{}^2(x_i) - \nu_A{}^2(x_i)\right|\right)\right] \tag{6.4.1}$$

Since $A \cap B = \{< x, (\mu_A(x) \wedge \mu_B(x)), (\nu_A(x) \vee \nu_B(x)) > | \, x \in X\}$

Thus $A \cap B = \{< x, (\mu_A(x)), (\nu_A(x)) > | \, x \in X_1\}$

and $A \cap B = \{< x, (\mu_B(x)), (\nu_B(x)) > | \, x \in X_2\}$

$$E(A \cap B) = \frac{1}{n}\sum_{i=1}^{n}\left[\cos\frac{\pi}{2}\left(\left|\mu_{A\cap B}{}^2(x_i) - \nu_{A\cap B}{}^2(x_i)\right|\right)\right]$$

$$=\frac{1}{n}\left[\sum_{x_i\in X_1}\cos\frac{\pi}{2}\left(\left|\mu_A^2(x_i)-\nu_A^2(x_i)\right|\right)+\sum_{x_i\in X_2}\cos\frac{\pi}{2}\left(\left|\mu_B^2(x_i)-\nu_B^2(x_i)\right|\right)\right]$$

(6.4.2.)

Adding (6.4.1) and (6.4.2), the expression becomes as:

$$E(A\cup B)+E(A\cap B)=$$

$$=\frac{1}{n}\left[\begin{array}{l}\sum_{x_i\in X_1}\cos\frac{\pi}{2}\left(\left|\mu_B^2(x_i)-\nu_B^2(x_i)\right|\right)+\sum_{x_i\in X_2}\cos\frac{\pi}{2}\left(\left|\mu_A^2(x_i)-\nu_A^2(x_i)\right|\right)\\+\sum_{x_i\in X_1}\cos\frac{\pi}{2}\left(\left|\mu_A^2(x_i)-\nu_A^2(x_i)\right|\right)+\sum_{x_i\in X_2}\cos\frac{\pi}{2}\left(\left|\mu_B^2(x_i)-\nu_B^2(x_i)\right|\right)\end{array}\right]$$

$$=\frac{1}{n}\sum_{i=1}^{n}\left[\cos\frac{\pi}{2}\left(\left|\mu_A^2(x_i)-\nu_A^2(x_i)\right|\right)\right]+\frac{1}{n}\sum_{i=1}^{n}\left[\cos\frac{\pi}{2}\left(\left|\mu_B^2(x_i)-\nu_B^2(x_i)\right|\right)\right]$$

$$=E(A)+E(B)$$

To test the consistency of the proposed entropy measure, numerical examples have been taken and discussed in the next section.

7. Numerical Examples

In this section, the comparison of the proposed entropy measure with the existing entropies proposed by (Ye 2010), (Verma & Sharma 2013), (Mishra 2016), (Wei 2012), (Wang & Wang 2012) and (Liu & Ren 2014) have been established. For an IFS A the notations: $E_{21}(A)$, $E_7(A)$, $E_6(A)$, $E_3(A)$, $E_4(A)$ and $E_5(A)$ stands for Ye, Verma & Sharma, Mishra, Wei, Wang & Wang and Liu & Ren entropy measure respectively.

The demonstrated is taken by means of two examples as:

Example I: The data has been taken from the (Mishra 2016) and is defined as:

Let $A=\{A_1, A_2, A_3, A_4\}$; where

$A_1 = <0.2, 0.5>$; $A_2 = <0.3, 0.5>$; $A_3 = <0.4, 0.5>$; $A_4 = <0.5, 0.5>$ be IFS in A

The comparison of the proposed entropy measure $E(A)$ with the other existing entropies are given in the following Table (1) as:

Table 1. Comparison of various entropy measures with the proposed entropy measure

IFSs	$E_{21}(A)$	$E_7(A)$	$E_6(A)$	$E_3(A)$	$E_4(A)$	$E_5(A)$	$E(A)$
A_1	0.9057	0.9138	0.8910	0.9056	0.6903	0.7146	**0.9461**
A_2	0.9580	0.9617	0.9511	0.9580	0.7673	0.7757	**0.9685**
A_3	0.9896	0.9903	0.9877	0.9896	0.8665	0.8676	**0.9900**
A_4	1.000	1.000	1.0000	1.000	1.000	1.000	**1.000**

From Table (1), it is observed that $E(A_4) \geq E(A_3) \geq E(A_2) \geq E(A_1)$

Therefore, the IF information measures reaches their maximum, when membership and non-membership values are equal, as given in example (1).

Example II: The data has been taken from (Liu & Ren 2014) and is defined as:

Let $A=\{A_1, A_2, A_3, A_4\}$; where

$A_1 = <0.4, 0.1>$; $A_2 = <0.6, 0.3>$; $A_3 = <0.2, 0.6>$; $A_4 = <0.13, 0.565>$ be the intuitionistic fuzzy sets, the values of the proposed entropy measure along with other entropy measures are given in the following Table (2) as:

Table 2. Comparison of various entropy measures with the proposed entropy measure

IFSs	$E_{21}(A)$	$E_7(A)$	$E_6(A)$	$E_3(A)$	$E_4(A)$	$E_5(A)$	$E(A)$
A_1	0.9057	0.9138	0.8910	0.9877	0.7265	0.7883	**0.9723**
A_2	0.9057	0.9138	0.8910	0.9771	0.6427	0.6457	**0.9114**
A_3	0.8329	0.8463	0.8090	0.9659	0.5774	0.5914	**0.8763**
A_4	0.8027	0.8180	0.7754	0.9659	0.5774	0.6103	**0.8893**

From Table (2), it is seen that $E(A_1) \geq E(A_2)$ and $E(A_3) \leq E(A_4)$.

Result and Discussion

From the tables (1-2), it is concluded that the proposed trigonometric IF entropy measures gives better insights of the available information as compared to the existing trigonometric entropy measures. The results obtained from the examples (I-II) explains the consistent nature of the proposed entropy measure and is better one against the existing entropies.

Conclusion

A new generalized information theoretic measure as trigonometric IFE measure has been proposed, which satisfies certain axioms required for a candidate measure to justify its validity. Further, numerical examples have been shown to compare the proposed entropy measure with the existing entropy measures. The proposed entropy measure has applications in decision-making across domains.

References

Ash, R. B. (1965). Information theory. *Dover Publications*, Inc, New York.

Atanassov, K. T. (1986). Intuitionistic fuzzy sets. *Fuzzy Sets and Systems*, 20(1), 87-96.

Atanassov, K. T. (1999). *Intuitionistic fuzzy sets.* Physica-Verlag Heidelberg - New York, 1-137.

Burillo, P. and Bustince, H. (1996). Entropy on intuitionistic fuzzy sets and on interval valued fuzzy sets. *Fuzzy Sets and Systems*, 78(3), 305-316.

Burillo, P., Bustince, H. and Mohedano, V. (1994). Some definition of intuitionistic fuzzy number. *Fuzzy based expert systems*, 28-30.

Chen, S. M. and Tan, J. M. (1994). Handling multi-criteria fuzzy decision-making problems based on vague set theory. *Fuzzy Sets and Systems*, 67(2), 163-172.

Das, S., Guha, D. and Dutta, B. (2016). Medical diagnosis with the aid of using fuzzy logic and intuitionistic fuzzy logic. *Applied Intelligence*, 45(3), 850- 867.

De Luca, A. and Termini, S. (1972). A definition of non-probabilistic entropy in the setting of fuzzy set theory. *Information and Control*, 20(4), 301-312.

De, S. K., Biswas, R. and Roy, A. R. (2001). An application of intuitionistic fuzzy sets in medical diagnosis. *Fuzzy Sets and Systems*, 117(2), 209-213.

De, S. K., Biswas, R. and Roy, A. R. (2003). Some Operations on Intuitionistic Fuzzy Sets. *Fuzzy Sets and Systems*, 114(3), 477–484.

Dubois, D. and Prade, H. (1980). Fuzzy sets and systems: Theory and Applications. *Academic Press*, New York, 49.

Fan, J. L. and Ma Y. L. (2002). Some new fuzzy entropy formulas. *Fuzzy sets and Systems*, 128(2), 277-284.

Gau, W. L. and Buehrer, D. J. (1993). Vague sets. *IEEE Transactions on Systems, Man, and Cybernetics*, 23, 610–614.

Havrda, J. and Charvát, F. (1967). Quantification method for classification processes: Concept of structural α –entropy'. *Kybernetika*, 3(1), 30-35.

Hooda, D. S. and Mishra, A. R. (2015). On trigonometric fuzzy information measures. *ARPN Journal of Science and Technology*, 5(3), 145-152.

Hung, W. and Yang, M. S. (2004). Similarity measures of intuitionistic fuzzy sets based on Hausdorff distance. *Pattern Recognition*, 25, 1603–1611.

Kumar, V., Arora, H., and Pal, Kiran (2016). Intuitionistic trapezoidal fuzzy prioritized weighted geometric operator: an algorithm for the selection of suitable treatment for lung cancer. *Proceedings of Fifth International Conference on Soft Computing for Problem Solving*, Volume 436 of the series Advances in Intelligent Systems and Computing, 605-615, Springer. Kumar, V. and Jain, S. (2018). Alternate procedure for the diagnosis of malaria via Intuitionistic Fuzzy sets, In: Panigrahi B., Hoda M., Sharma V., Goel S. (eds). *Nature Inspired Computing*, Advances in Intelligent Systems and Computing, 652, 49-53, Springer, Singapore.

Liu, H. W. and Wang, G. J. (2007). Multi-criteria decision-making methods based on intuitionistic fuzzy sets. *European Journal of Operational Research*, 179, 220–233.

Liu, M. and Ren, H. (2014). A new intuitionistic fuzzy entropy and application in multi-attribute decision-making. *Information*, 5, 587-601.

Mishra, A. R. (2016). Intuitionistic fuzzy information measures with application in rating of township development. *Iranian Journal of Fuzzy system*, 13(3), 49-70.

Pal, N. R. and Pal, S. K. (1989). Object background segmentation using new definitions of entropy. *IEEE Proc*. E, 136(4), 284–295.

Parkash, O., Sharma, P. K. and Mahajan, R. (2008). New measures of weighted fuzzy entropy and their applications for the study of maximum weighted fuzzy entropy principle. *Information Sciences*, 178(11), 2839-2395.

Rényi, A. (1961). On measures of entropy and information. *Proceedings of the Fourth Berkeley Symposium on Mathematical Statistics and Probability,* Berkeley, CA: University of California Press, 547-561.

Sanchez, E. (1977a). Solutions in composite fuzzy relation equation: an application to medical diagnosis. *Fuzzy Automata and Decision Process*, Elsevier.

Sanchez, E. (1977b). Truth-qualification and fuzzy relations in natural languages, application to medical diagnosis. *Fuzzy Sets and Systems*, 84, 155–167.

Shannon, C. E. (1948). A mathematical theory of communication. *The Bell Syst Tech Journal*, 27(3), 379-423.

Szmidt, E. and Kacprzyk, J. (2001a). Entropy for intuitionistic fuzzy sets. *Fuzzy Sets and Systems*, 118(3), 467-477.

Szmidt, E. and Kacprzyk, J. (2001b). Distances between intuitionistic fuzzy sets. *Fuzzy Sets Syst*, 114(3), 505–518.

Szmidt, E. and Kacprzyk, J. (2001c). Intuitionistic fuzzy sets in some medical applications. *Computational Intelligence: Theory and Applications (B. Reusch, Ed.)* Springer, Berlin, 148-151.

Tsallis, C. (1988). Possible generalization of Bolzmann-Gibbs statistics. *J. Stat. Phys.*, 52, 479-487.

Verma, R. and Sharma, B. D. (2013). Exponential entropy on intuitionistic fuzzy sets. *Kybernetika*, 49, 114–127.

Vlachos, I. K. and Sergiadis, G. D. (2007). Intuitionistic fuzzy information- application to pattern recognition. *Pattern Recognition Letters*, 28(2), 197–206.

Wang, J. and Wang, P. (2012). Intuitionistic linguistic fuzzy multi-criteria decision- making method based on intuitionistic fuzzy entropy. *Control Decis.*, 27, 1694–1698.

Wang, J. Q. and Zhang, Z. H. (2009a). Multi-criteria decision-making method with incomplete certain information based on intuitionistic fuzzy number. *Control and Decision*, 24, 226-230.

Wang, J. Q. and Zhang, Z. H. (2009b). Aggregation operators on intuitionistic trapezoidal fuzzy number and its application to multi-criteria decision- making problems. *Journal of Systems Engineering and Electronics*, 20, 321-326.

Wei, C., Gao, Z. and Guo, T. (2012). An intuitionistic fuzzy entropy measure based on trigonometric function. *Control Decis.*, 27, 571–574.

Wei, G. W. (2012). Hesitant fuzzy prioritized operators and their application to multiple attribute decision-making. *Knowledge-Based Systems*, 31, 176-182.

Xu, Z. S. and Yager, R. R. (2006). Some geometric aggregation operators based on intuitionistic fuzzy sets. *International Journal of General Systems*, 35, 417–433.

Yazdani, M., Chatterjee, P., Zavadskas, E. K. and Hashemkhani Zolfani, S. (2017). Integrated QFD-MCDM framework for green supplier selection. *Journal of Cleaner Production*, 142, 3728–3740.

Ye, J. (2010). Two effective measures of intuitionistic fuzzy entropy. *Computing*, 87(1-2), 55-62.

Zadeh, L. A. (1965). Fuzzy sets. *Info. Contr.*, 8, 338-353.

Zadeh, L. A. (1968). Probability measure of fuzzy events. *Journal of Mathematical Analysis and Applications*, 23(2), 421-427.

Zadeh, L. A. (1969). Biological applicationof the theory of fuzzy sets and systems. *In: The Proceedings of an International Sysposiumon Biocybernatics of central Nervous System. Boston: Little, Brown and Company*, 199-206.

Zamanzade, E. (2014). Testing uniformity based on new entropy estimators. *Journal of Statistical Computation and Simulation*, 85(16), 3191-3205.

Zamanzade, E. and Arghami, N. R. (2011). Goodness-of-fit test based on correcting moments of modified entropy estimator. *Journal of Statistical Computation and Simulation*, 81(12), 2077-2093.

Zamanzade, E. and Arghami, N. R. (2012). Testing normality based on new entropy estimators. *Journal of Statistical Computation and Simulation*, 82(11), 1701-1713.

Zamanzade, E. and Mahdizadeh, M. (2016). Entropy estimation from judgement post stratified data. *Pakistan Journal of Statistics and Operation Research*, 7(4), 625-637.

Zamanzade, E. and Mahdizadeh, M. (2017). Entropy estimation from ranked set samples with application to test of fit. *Revista Colombiana de Estadstica*, 40(2), 223-241.

Zhang, Q. S. and Jiang, S. Y. (2008). A note on information entropy measure for vague sets. *Information Sciences*, 178(21), 4184-4191.

Chapter 6

Parametric Intuitionistic Fuzzy Entropies with Applications in Insurance

Prince Goyal*

Department of Community Medicine, Dr. Yashwant Singh Parmar Govt. Medical College, Nahan, Distt. Sirmour, (Himachal Pradesh), India

Abstract

Life is beautiful but also uncertain and there are many requirements to secure our life. Depending on the need, a suitable insurance policy that provides financial security to the family must be selected. In this chapter, an attempt has been made to propose the solution for the selection of suitable life insurance policy. Based on the requirements, a suitable insurance policy can be offered to the customers. One parametric entropy measure in an intuitionistic fuzzy environment has been proposed to study the problem of the selection of suitable insurance policy. For this purpose, a case study containing information about the customers, their requirements, and the type of available life insurance policies with assigned degree of membership, non-membership and intuitionistic index has been considered for the study.

Keywords: intuitionistic fuzzy sets, parametric entropies, Tsallis entropy measure, Rényi's entropy measure, decision-making

* Corresponding Author's Email: princegoyel@gmail.com.

In: Understanding Information Entropy
Editor: Vijay Kumar
ISBN: 979-8-88697-839-1

1. Introduction

Decision-making is the most popular and common activity of human beings. It is an essential part of our daily life activities. Generally, in any decision-making process, a set of alternatives are given with a variety of actions in which suitable alternative will be chosen. In the 18th century, the word Decision-Making was used first time in France for the methods of elections and social choice. A decision-making problem can also be known as an optimization problem. The complete knowledge of decision-making prepares a person for taking better decisions under risk and also predicts the behavior of others. Moreover, this process codifies our decision-making methodology in a mathematically tractable format. Decision-making techniques have lots of applications across domains, including investments, medical diagnosis and in routine work. Investments are the backbone of any economy and business. It is an art of committing money to venture while expecting some extra benefits. In this growing market, the investor is truly the emperor. If the investor has some knowledge of the nature of the market, he can gain fruitful and profitable returns. During the last few decades, a peculiar and remarkable change has been recorded in financial and economic infrastructure across the globe. So, investing in his own assets is not an easy task for a customer when various options are available in the market.

Life is admirable but full of risk and vagueness. Whatever we do, irrespective of how smart and hard, we are never sure what life has in store for us. Therefore, it is important for us to secure our lives financially by investing in some venture. Life insurance is one such type of investment in which we not only secure our family but also secure our future. Life Insurance can be defined as a contract between the policy holder and the insurance company, where the life insurance company pays a specific amount of money to the insured individual's family upon his death or after a fixed time period. Therefore, it is important to get better insurance plan that satisfies the requirements. It is important to compare the life insurance policies and select the best policy offered by different companies according to their contract terms, cost premium quotes, limitations and benefits. For better results in the selection of insurance policy, various types of insurance plans and their features have been defined in the IF environment. A hypothetical case study has been demonstrated in this chapter that contains information about customers, their requirements, the types of life insurance policies with an assigned degree of membership, non-membership and intuitionistic index.

2. Fuzzy Sets

The world is full of situations that do not have defined boundaries, which can be handled by the concept of fuzzy theory. In 1965, Prof. Zadeh, gave the concept of fuzzy set theory, which is the generalization of classical set theory. Shannon (1948) introduced the concept of information theory, which is indeed a strong tool to manipulate data and information possessed uncertainty.

A fuzzy set A is defined in a discrete universe of discourse $X=\{x_1,x_2,...,x_n\}$ as: $A=\{<x,\ \mu_A(x)>\ :\ x\in X\}$ where,

$\mu_A : X \to [0,1]$ is the membership function of set A and

$\mu_A(x)$ is called the grade of membership of $x \in X$ in A

De Luca and Termini [1972] gave the fuzzy measure corresponding to the Shannon's measure. Later on Kapur [1997], Hooda and Jain [2009], Zarandi [2010], Deshmukh et al. [2011], Kumar et al. [2011], Ryjov [2009], Verma and Sharma [2011, 2014a], Adlassnig [1980, 1982], Bhandari and Pal [1993], etc. proposed so many fuzzy measures.

3. Intuitionistic Fuzzy Set Theory

Atanassov (1986) An IFS A in $X=\{x_1,x_2,...,x_n\}$ is given as:

$A=\{<x,\ \mu_A(x),\ \nu_A(x)>\mid x\in X\}$; where

$\mu_A : X \to [0,1]$ and $\nu_A : X \to [0,1]$ are the membership and non-membership functions of set A.

$\pi_A(x)=1-\mu_A(x)-\nu_A(x)$ is intuitionistic index or hesitation index of x in A.

In limiting case, if $\pi_A(x)=0$, IFS reduces automatically to FS.

Some properties of IFS theory are:

Let $A, B \in X$ be two IFS s.t.

$A=\{<x,\ \mu_A(x),\ \nu_A(x)>\mid x\in X\}$, $B=\{<x,\ \mu_B(x),\ \nu_B(x)>\mid x\in X\}$

Atanassov (1999) and (Szmidt & Kacprzyk 2001a), operations on IFS are given as:

(a) $A \subseteq B$ *iff* $\mu_A(x) \le \mu_B(x)$ and $\nu_A(x) \ge \nu_B(x)\ \forall\, x \in X$;

(b) $A = B$ *iff* $A \subseteq B$ and $B \subseteq A$;

(c) $A^C = \{< x, \nu_A(x), \mu_A(x) > | \, x \in X\}$;

(d) $A \cup B = \{< x, (\mu_A(x) \vee \mu_B(x)), (\nu_A(x) \wedge \nu_B(x)) > | \, x \in X\}$;

(e) $A \cap B = \{< x, (\mu_A(x) \wedge \mu_B(x)), (\nu_A(x) \vee \nu_B(x)) > | \, x \in X\}$;

(f) $A @ B = \left\{< x, \dfrac{\mu_A(x) + \mu_B(x)}{2}, \dfrac{\nu_A(x) + \nu_B(x)}{2} > | \, x \in X\right\}$;

(g) $A \le B$ *iff* $\mu_A(x) \le \mu_B(x)$ and $\nu_A(x) \ge \nu_B(x)$ for $\mu_B(x) \le \nu_B(x)$

or $\mu_A(x) \ge \mu_B(x)$ and $\nu_A(x) \le \nu_B(x)$ for $\mu_B(x) \ge \nu_B(x)$

$\forall\, x \in X$.

The work in this direction has been proposed by Atanassov [1994, 1994a], Cagman and Karatas [2013], Hung and Yang [2006], Ejewa et al. [2014, 2014a], Cornelis et al. [2004], Husain et al. [2010, 2012], Ahmed et al. [2010], Kumar [2011, 2013], Prince et al. [2015], etc. Neog and Sut [2011] applied fuzzy soft sets and fuzzy soft matrices to decision-making problems. Chen [1995] defined similarity measures between vague sets. Maji and Roy [2002] have used soft sets in decision-making. Szmidt and Kacprzyk [2001, 2001a, 2001b, 2003] built some soft decision-making models and entropies in the IF environment. Liu [2005] derived new similarity measures between elements and IF sets. Jha and Verma [2012] discussed some new measures of IF entropy.

In section (4) the representation of IF measures has been defined graphically. Section (5) deals with some entropy measures for IFSs. In section (6) Rényi's and Tsallis entropies (one parametric) in IF environment have been used. Various life insurance policies have been discussed in section (7). Finally, in section (8), a hypothetical case study for the selection of a better life insurance policy using Rényi's and Tsallis parametric intuitionistic fuzzy measures has been demonstrated. Later on, the conclusion and future aspects are discussed.

4. Graphically Representation of an Intuitionistic Fuzzy Measure

The graphical representation is given as:

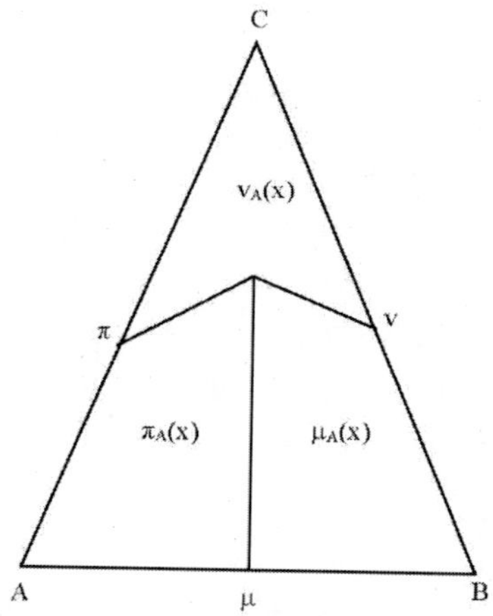

Figure 1. Graphical Representation of an Intuitionistic Fuzzy Measure.

5. Some Entropy Measures for IFSs

De Luca and Termini [1972] were the first to introduce the concept of non-probabilistic entropies with various properties such as sharpness, maximality resolution and symmetry.

Burillo and Bustince [1996, 1996a] first axiomized theory of entropy for IF sets and are defined as:

A real function $e: FS(X) \to \Re^+$ is called entropy in a Fuzzy Set, if it holds the following properties.

(a) $e(A) = 0$, If A is a crisp set

(b) $e(A) = cardinal(X)$ *iff* $\mu_A(x) = \nu_A(x) = 0$; $\forall\, x \in FS(X)$

(c) $e(A)$ attains a unique maximum, if $\mu_A(x) = \frac{1}{2}$

(d) $e(A) \geq e(B)$ *iff* $\mu_A(x) \leq \mu_B(x)$ and $\nu_A(x) \leq \nu_B(x)$; $\forall\, x \in FS(X)$

(e) $e(A) \geq e(A^c)$; $\forall\, x \in FS(X)$

where, $e(A)=\sum_{i=1}^{|x|}\pi_A(x_i)$ and | x | denotes the cardinality of X.

Again, Szmidt and Kacprzyk [2001b] proposed distance measures and some entropy functions in IF environment as:

$$e(A)=\frac{1}{|X|}\sum_{i=1}^{|x|}\frac{\min\{\mu_A(x_i),\nu_A(x_i)\}+\pi_A(x_i)}{\max\{\mu_A(x_i),\nu_A(x_i)\}+\pi_A(x_i)}$$

Hung and Yang [2006] introduced an IF measure by using π-function (hesistant index) and proposed that entropy is maximum when all the three characters are equal and minimum when only one of the characters is present and rest are zero, defined as:

A real function $E: IFS(X)\to\Re^+$ is called entropy on IF Set, if it holds the following properties as:

(a) $E(A)=0$, If A is a crisp set

(b) $E(A)$ attains unique maximum, if $\mu_A(x)=\nu_A(x)=\pi_A(x)=\frac{1}{3}$

(c) $E(A)\le E(B)$ *iff* $\mu_A(x)\le\mu_B(x)$ and $\nu_A(x)\le\nu_B(x)$

for $\max\{\mu_B(x),\nu_B(x)\}\le\frac{1}{3};\forall\, x\in IFS(X)$

and $\mu_A(x)\ge\mu_B(x)$ and $\nu_A(x)\ge\nu_B(x)$ for $\min\{\mu_B(x),\nu_B(x)\}\ge\frac{1}{3}$;

$\forall\, x\in IFS(X)$

(d) $E(A)=E(A^c)$; $\forall\, x\in FS(X)$

Szmidt and Kacprzyk [2000] described measures for IF sets as follows:

$$d_H(A,B)=\sum_{i-1}^{N}\left(|\mu_A(x_i)-\mu_B(x_i)|+|\nu_A(x_i)-\nu_B(x_i)|+|\pi_A(x_i)-\pi_B(x_i)|\right);$$

$$0\le d_H(A,B)\le N$$

Also, Wang and Xin [2005] gave distance measures between two IF sets as:

$$d(A,B)=\frac{1}{N}\sum_{i=1}^{N}\left(\frac{|\mu_A(x_i)-\mu_B(x_i)|+|\nu_A(x_i)-\nu_B(x_i)|}{4}+\max\left[\frac{|\mu_A(x_i)-\mu_B(x_i)|,|\nu_A(x_i)-\nu_B(x_i)|}{2}\right]\right)$$

and $d(A,B)=\frac{1}{\sqrt[l]{N}}\left[\sum_{i=1}^{N}\left(\frac{|\mu_A(x_i)-\mu_B(x_i)|}{2}+\frac{|\nu_A(x_i)-\nu_B(x_i)|}{2}\right)\right]^{\frac{1}{l}}$

Granstrom [2011] and Atanassov [1999, 2012] proposed the theoretical aspects of the IFSs and apply them in various sectors. Szmidt [2014] derived similarities and distances in IFSs. Mints [2002] introduced the concept of intuitionistic logic. Prince et al. [2015] derived some IF measures which have applications in decision-making problems. Zhao et al. [2010] used the concept of aggregation operators in IF set theory.

6. Rényi's and Tsallis Entropies in Intuitionistic Fuzzy Environment

Both entropies (Rényi's and Tsallis) are generalizations of Shannon's entropy. These entropies have been extensively studied in the literature from an axiomatic point of view. These are unique under some natural assumptions, such as additivity, α-additivity, etc.

Rényi [1961] proposed a generalized entropy measure which is defined as:

$$R_\alpha(P)=\frac{1}{1-\alpha}\log\left[\sum_{i=1}^{n}p_i^\alpha\right], \qquad \alpha\neq 1,\alpha>0$$

Some properties of the measure are:

1. Rényi's entropy is a non-increasing convex function.
2. This entropy is decisive.

3. If $\alpha \le 1$ then Rényi's entropy is concave. If $\alpha > 1$ the entropy is neither pure convex nor pure concave.
4. Rényi's entropy is also non-negative.
5. The entropy function is bounded and continuous for α.
6. Rényi's entropy for different values of α is calculated.

Tsallis [1988] proposed a generalized entropy function as:

$$H_{\alpha}^{T}(P) = \frac{1}{1-\alpha}\left[\sum_{i=1}^{n} p_i^{\alpha} - 1\right], \qquad \alpha \neq 1, \alpha > 0$$

Some properties of the Tsallis entropy as:

1. Tsallis entropy is strictly convex.
2. This entropy is non-additive.
3. Tsallis entropy is uniquely determined by a small number of axioms.
4. Tsallis entropy is monotonically increasing and continuous.
5. Tsallis entropy also contains the property of involving conditional entropies.

Jha and Verma [2012] proposed IF version of Rényis and Tsallis entropies as follows:

$$F_1(A) = \frac{1}{1-\alpha}\log\sum_{i=1}^{n}\left[\mu_A^{\alpha}(x_i) + v_A^{\alpha}(x_i)\right], \qquad \alpha > 0, \alpha \neq 1$$

and

$$F_2(A) = \frac{1}{1-\alpha}\sum_{i=1}^{n}\left[\mu_A^{\alpha}(x_i) + v_A^{\alpha}(x_i) - 1\right], \qquad \alpha > 0, \alpha \neq 1$$

7. Life Insurance Policy and Its Types

Human life is precious and important, but risks and uncertainties are the unavoidable parts. If some unfortunate incident happens, then a question arises about who will provide financial security and safety for the family? Life insurance provides a way for our family to continue their living comfortably.

Therefore, it is necessary to select a good insurance plan with a desired investment portfolio. There are so many life insurance policies available in the market offered by different companies and it is very difficult to analyze the best policy among the available ones. The different types of insurance policies are defined as:

7.1. Term Insurance

Term insurance is a traditional type of life insurance policy that gives advantages to a certain and specific period of time. If the life insured dies within the specific time period, then death benefit will be given. Term insurance plans generally offer financial security for the whole family in case of the unfortunate death of the policyholder. Under this, various other benefits can be availed, such as critical illnesses or accidental death. But the premium of such policies is very high and depends upon the duration of the policy. These insurance plans are more reasonable than money back policies, endowment policies and ULIPs. A term life insurance policy is the simplest and cheapest type of life insurance. Policy renewable, convertibility, changes in premiums, decrease in terms are the main features of term life insurance policies.

7.2. Whole Life Insurance Policy

The whole life insurance policy includes both insurance and an investment component. Whole life insurance is a contract that provides insurance coverage for entire life and gives a certain amount upon the death of the insured. This insurance policy is a saving scheme and also has cash value or loan value. Whole life insurance lasts for an insured's lifetime, as opposed to term life insurance, which is for a specific number of years. Whole life insurance is paid out to the beneficiary or beneficiaries even after the insured's death, provided that the policy was in force. Whole life insurance is also called ordinary life insurance, permanent life insurance or straight life insurance.

7.3. Endowment Policy

Endowment policy pays the assured sum at expiration of a fixed date or upon the death of the policy holder, whichever comes earlier. The charity that gets a person's money a year after his death is an example of an endowment policy. The premium of endowment policy may be higher than term insurance and whole life policies. An endowment is also called endowment life policy and provides ongoing benefits. In some endowment plans, a fixed percentage of the assets is allowed to be used, which may be a combination of interest income and principal. The unique feature of endowment plans is that they provide benefits upon maturity. On the completion of the tenure of the policy, the policyholder receives a maturity benefit and premium payment. Usually, we can select the premium payment as per our comfort.

7.4. Money Back Policy

Money back policies are also called anticipated endowment plans and more popular than other life insurance policies, according to the market survey. In the money back policy, the policy holder gets a percentage of the whole money as a survival benefit, in regular installments, instead of getting the lump sum amount at the time of maturity. If the insured dies before the maturity period or during the policy period, the nominee gets the whole sum assured and the policy terminates. The money back policy is suitable for risk average gives both maturity benefit and death benefit. Money back policy is an endowment plan with the benefit of liquidity. The main features of the money back policy are: Guaranteed returns from the money back policy, Income during the lifetime of the money back plan, Income on the maturity of the plan, Income on the death of the insured person in the policy income. Bonus amounts also helps in increasing the money back policy, etc.

7.5. Unit Linked Insurance Plans

ULIP acts just like a saving vehicle that includes insurance cover and investment. In ULIPs, some part of the installment is utilized to provide risk cover for the insured while the remaining portion is invested in various markets schemes (equity and debt schemes). There are so many ULIPs available in the market, such as retirement plans, health plans, children's education plans and some other investment plans. The units or funds have a net asset value (NAV) that changes according to the performance of the market, which is allotted to policy holders. The main features of ULIP are flexibility, life protection, market linked returns, risk saving coverage, etc.

Thus, one can say that ULIP is a pure type of insurance policy that gives advantage of both insurance and investment in one plan.

8. Selection of Life Insurance Policy using Entropy Measures in IF Environment

Investment including gold, property, stock market, fixed income instruments, mutual funds and, of course, life insurance is the best option. Given the plethora of choices, it becomes imperious to make the right choice while investing our hard-earned money. There are many types of life insurance policies, like term insurance policy, whole life policy, endowment policy, money back policy, unit linked insurance plan, universal life insurance, permanent life insurance, etc. with features such as risk cover, tax benefit, investment, loan facility, exchange of policy, mortgage recovery, etc. The requirements of the customers (policy holders) are changing with the passage of time and the selection of a suitable insurance plan with specific needs is a million dollar question for everyone. In this chapter, an IF information technique has been proposed in which Max-Min-Max IF relations with one parametric entropy measure like Rényi's and Tsallis measures has been utilized to select the best insurance policy.

Yadav and Tiwari [2012] studied factors that affect customer's investment in life insurance policies. Kothari et al. [2013] studied factors affecting individual's investment towards life insurance policies. Dash and Sood [2013] discussed why a person should invest in a life insurance policy. Upadhyay [2013] described satisfaction of the policy holders' protection in the insurance sector. Sharma and Chowhan [2013] did a comparative study of public and private life insurance companies. Barik and Patra [2014] have done empirical research on emerging trends in insurance in India. Horn [1971] also explained life insurance earnings and release from the risk policy reserve system.

8.1. To Choose Insurance Policy

In this case study, exploratory and descriptive type research was carried out to describe the phenomenon in IF set theory. The study aims to find out that type of life insurance policy which is more suitable for a customer by studying factors influencing customer's life insurance investment decision and their preferences at the time of the policy buying decision. In this procedure: Firstly,

IF relation between customers and features has been established that affects a life insurance policy with assigned degree of membership and non-membership elements given in Table (1). Also, IF relation between features and types of life insurance policies has been established with assigned degree of membership and non-membership elements as given in Table (2) by taking information about different insurance policies. In Table (3), max-min composition of IF relations has been used. Finally, Rényi's entropy and Tsallis's entropy in IF environment for different values of the parameter (α) has been used and given in Tables (4a-4i) and (5a-5i). The minimum weight for each customer from the possible analysis is the decision value. Moreover, if there is a tie in weight elements, then there may be a chance that the customer can purchase both types of life insurance policies.

Let $F=\{f_1,....,f_m\}$, $I=\{i_1,....,i_n\}$ and $C=\{c_1,....,c_q\}$ be the finite set of features, types of life insurance policies and customers respectively.

Assuming that $\mu_Q(c,f)$ is the degree to which the customer c is affected by the feature f and $\nu_Q(c,f)$ indicate the degree to which the customer c is not affected by the feature f.

Similarly, $\mu_R(f,i)$ represents the degree to which the feature f affects the type of life insurance policy i and $\nu_R(f,i)$ represents the degree to which the feature f does not affects the type of life insurance policy i.

The composition $T(R\circ Q)$ of IF relations T and Q describe the interest of customer c_i in terms of choosing the life insurance policy from C to I given by membership and non-membership as:

$$\mu_T(c_i,i)=\max_{f\in F}\{\min[\mu_Q(c_i,f),\mu_R(f,i)]\}$$

$$\nu_T(c_i,i)=\min_{f\in F}\{\max[\nu_Q(c_i,f),\nu_R(f,i)]\};\forall c_i\in C \quad \text{and} \quad i\in I$$

The IF version of Rényi's and Tsallis entropies are defined as:

$$R_\alpha(A) = \frac{1}{1-\alpha} \log \sum_{i=1}^{n} \left(\mu_i^\alpha + \nu_i^\alpha + \pi_i^\alpha \right); \alpha \neq 1, \alpha > 0$$

and

$$H_\alpha^T(A) = \frac{1}{1-\alpha} \left[\sum_{i=1}^{n} \left(\mu_i^\alpha + \nu_i^\alpha + \pi_i^\alpha \right) - 1 \right]; \alpha \neq 1, \alpha > 0$$

There are lots of features that affect the interest of a customer, like risk covered, investment, maturity benefits, saving component, cash value plan, tax benefits, loan facilities, mortgage protection, death benefits, health protection, etc. In our study, five factors (risk covered, investment, tax benefits, health protection, loan facilities) have been considered. Life insurance policies such as whole life policy, term insurance policy, limited payment life policy, fixed term endowment policy, money back policy, unit linked insurance plans, annuities, etc. have been studied and out of which five types of policies (term insurance, whole life, endowment, money back, unit linked insurance) have been considered for the study.

Let $C = \{c_1, ..., c_4\}$ be the number of customers and $F = \{f_1, ..., f_5\}$ be the set of features that affect the interest of a customer. Let $IFS : Q(C \to F)$ is given as:

Table 1. Membership and non-membership values of customers and features of a life insurance policy

Q	**F₁**		**F₂**		**F₃**		**F₄**		**F₅**	
Customer	**uQ**	**vQ**	**uQ**	**vQ**	**uQ**	**vQ**	**uQ**	**vQ**	**uQ**	**vQ**
C_1	0.7	0.2	0.6	0.0	0.1	0.7	0.5	0.2	0.0	0.6
C_2	0.0	0.6	0.5	0.3	0.6	0.1	0.0	0.6	0.0	0.9
C_3	0.5	0.2	0.7	0.2	0.0	0.5	0.2	0.3	0.3	0.4
C_4	0.4	0.0	0.3	0.5	0.2	0.3	0.8	0.1	0.5	0.3

Let $I = \{i_1, ..., i_5\}$ be the types of life insurance policies that a customer wants to purchase. Suppose $IFS : R(F \to I)$ is given as:

Table 2. Membership and non-membership values between features and types of a life insurance policy

R	I_1		I_2		I_3		I_4		I_5	
Factors	u_R	v_R	u_R	v_R	u_R	v_R	u_R	v_R	u_R	v_R
F_1	0.7	0.1	0.3	0.1	0.4	0.4	0.2	0.6	0.2	0.7
F_2	0.4	0.5	0.8	0.0	0.5	0.3	0.3	0.3	0.0	0.6
F_3	0.1	0.5	0.2	0.7	0.6	0.2	0.5	0.3	0.2	0.4
F_4	0.3	0.2	0.5	0.3	0.6	0.1	0.7	0.2	0.4	0.3
F_5	0.2	0.5	0.1	0.6	0.2	0.7	0.3	0.6	0.9	0.0

The composition $T = R \circ Q$ is given as:

Table 3. Membership and non-membership values of customers and their corresponding types of life insurance policies

T	I_1		I_2		I_3		I_4		I_5	
Customers	u_Q	v_Q	u_Q	v_Q	u_Q	v_Q	u_Q	v_Q	u_Q	v_Q
C_1	0.7	0.2	0.6	0.0	0.1	0.7	0.5	0.2	0.0	0.6
C_2	0.4	0.5	0.5	0.3	0.6	0.2	0.5	0.3	0.2	0.4
C_3	0.5	0.2	0.7	0.2	0.5	0.3	0.3	0.3	0.3	0.3
C_4	0.4	0.1	0.5	0.1	0.6	0.1	0.7	0.2	0.5	0.3

Using Rényi's and Tsallis measures of entropies, the results are given in the following tables as:

For Rényi's Entropy

The values are:

Table 4a. when $\alpha = 0.1$

R(p)	I_1	I_2	I_3	I_4	I_5
C_1	0.463	**0.300**	0.463	0.474	**0.300**
C_2	**0.468**	0.474	0.471	0.474	0.475
C_3	0.474	**0.463**	0.474	0.476	0.476
C_4	0.468	0.468	0.466	**0.463**	0.474

Table 4b. when $\alpha = 0.2$

R(p)	**I_1**	**I_2**	**I_3**	**I_4**	**I_5**
C_1	0.449	**0.239**	0.449	0.471	0.239
C_2	**0.460**	0.471	0.464	0.471	0.472
C_3	0.471	**0.449**	0.471	0.476	0.476
C_4	0.460	0.461	0.456	**0.449**	0.471

Table 4c. when $\alpha = 0.3$

R(p)	**I_1**	**I_2**	**I_3**	**I_4**	**I_5**
C_1	0.435	**0.298**	0.435	0.468	**0.298**
C_2	**0.452**	0.468	0.458	0.468	0.471
C_3	0.468	**0.435**	0.468	0.476	0.476
C_4	0.452	0.452	0.448	**0.435**	0.468

Table 4d. when $\alpha = 0.4$

R(p)	**I_1**	**I_2**	**I_3**	**I_4**	**I_5**
C_1	0.421	**0.297**	0.421	0.464	0.297
C_2	**0.444**	0.464	0.451	0.464	0.468
C_3	0.464	**0.421**	0.464	0.475	0.475
C_4	0.444	0.444	0.437	**0.421**	0.464

Table 4e. when $\alpha = 0.5$

R(p)	**I_1**	**I_2**	**I_3**	**I_4**	**I_5**
C_1	0.408	**0.296**	0.408	0.462	**0.296**
C_2	**0.438**	0.462	0.445	0.462	0.467
C_3	0.462	**0.408**	0.462	0.475	0.475
C_4	0.438	0.438	0.428	**0.408**	0.462

Table 4f. when $\alpha = 0.6$

R(p)	**I_1**	**I_2**	**I_3**	**I_4**	**I_5**
C_1	0.395	**0.295**	0.395	0.458	**0.295**
C_2	**0.431**	0.458	0.438	0.458	0.465
C_3	0.458	**0.395**	0.458	0.474	0.474
C_4	0.431	0.431	0.420	**0.395**	0.458

Table 4g. when $\alpha = 0.7$

R(p)	I_1	I_2	I_3	I_4	I_5
C_1	0.382	**0.294**	0.382	0.455	**0.294**
C_2	**0.425**	0.455	0.431	0.455	0.463
C_3	0.455	**0.382**	0.455	0.474	0.474
C_4	0.425	0.425	0.412	**0.382**	0.455

Table 4h. when $\alpha = 0.8$

R(p)	I_1	I_2	I_3	I_4	I_5
C_1	0.371	**0.294**	0.371	0.453	**0.294**
C_2	**0.420**	0.453	0.425	0.453	0.462
C_3	0.453	**0.371**	0.453	0.474	0.474
C_4	0.420	0.420	0.404	**0.371**	0.453

Table 4i. when $\alpha = 0.9$

R(p)	I_1	I_2	I_3	I_4	I_5
C_1	0.359	**0.293**	0.359	0.450	**0.293**
C_2	**0.414**	0.450	0.419	0.450	0.459
C_3	0.450	**0.359**	0.450	0.473	0.473
C_4	0.414	0.414	0.397	**0.359**	0.450

From the above given tables, it is concluded that customer C_2 can purchase life insurance policy I_1 , customer C_4 can purchase insurance policy I_4 where as the customers C_1 and C_3, can opt for life insurance policy I_2.

For Tsallis Entropy

The values are:

Table 5a. when $\alpha = 0.1$

$H_\alpha(p)$	I_1	I_2	I_3	I_4	I_5
C_1	1.789	**0.958**	1.789	1.856	**0.958**
C_2	**1.822**	1.856	1.836	1.856	1.862
C_3	1.856	**1.789**	1.856	1.872	1.872
C_4	1.822	1.822	1.812	**1.789**	1.856

Table 5b. when $\alpha = 0.2$

$H_\alpha(p)$	**I_1**	**I_2**	**I_3**	**I_4**	**I_5**
C_1	1.060	**0.919**	1.608	1.726	**0.919**
C_2	**1.667**	1.726	1.690	1.726	1.737
C_3	1.726	**1.608**	1.726	1.755	1.755
C_4	1.667	1.667	1.649	**1.608**	1.726

Table 5c. when $\alpha = 0.3$

R(p)	**I_1**	**I_2**	**I_3**	**I_4**	**I_5**
C_1	1.452	**0.882**	1.452	1.608	**0.882**
C_2	**1.532**	1.608	1.559	1.608	1.623
C_3	1.608	**1.452**	1.608	1.647	1.647
C_4	1.532	1.532	1.508	**1.452**	1.608

Table 5d. when $\alpha = 0.4$

R(p)	**I_1**	**I_2**	**I_3**	**I_4**	**I_5**
C_1	1.317	**0.847**	1.317	1.501	**0.847**
C_2	**1.415**	1.501	1.443	1.501	1.519
C_3	1.501	**1.317**	1.501	1.547	1.547
C_4	1.415	1.415	1.385	**1.317**	1.501

Table 5e. when $\alpha = 0.5$

R(p)	**I_1**	**I_2**	**I_3**	**I_4**	**I_5**
C_1	1.200	**0.814**	1.200	1.404	**0.814**
C_2	**1.311**	1.404	1.338	1.404	1.424
C_3	1.404	1.200	1.404	1.455	1.455
C_4	1.311	1.311	1.277	**1.200**	1.404

Table 5f. when $\alpha = 0.6$

R(p)	**I_1**	**I_2**	**I_3**	**I_4**	**I_5**
C_1	1.098	**0.782**	1.098	1.315	**0.782**
C_2	**1.220**	1.315	1.243	1.315	1.337
C_3	1.315	**1.098**	1.315	1.370	1.370
C_4	1.220	1.220	1.182	**1.098**	1.315

Table 5g. when $\alpha = 0.7$

R(p)	**I_1**	**I_2**	**I_3**	**I_4**	**I_5**
C_1	1.009	**0.753**	1.009	1.234	**0.753**
C_2	**1.138**	1.234	1.158	1.234	1.257
C_3	1.234	**1.009**	1.234	1.291	1.291
C_4	1.138	1.138	1.098	**1.009**	1.234

Table 5h. when $\alpha = 0.8$

R(p)	**I_1**	**I_2**	**I_3**	**I_4**	**I_5**
C_1	0.930	**0.724**	0.930	1.159	**0.724**
C_2	**1.066**	1.159	1.082	1.159	1.184
C_3	1.159	**0.930**	1.159	1.219	1.219
C_4	1.066	1.066	1.023	**0.930**	1.159

Table 5i. when $\alpha = 0.9$

R(p)	**I_1**	**I_2**	**I_3**	**I_4**	**I_5**
C_1	0.862	**0.698**	0.862	1.091	**0.698**
C_2	**1.001**	1.091	1.012	1.091	1.116
C_3	1.091	**0.862**	1.091	1.151	1.151
C_4	1.001	1.001	0.957	**0.862**	1.091

From these tables, it is concluded that the decision values are same as received from Rényis entropy.

Conclusion

In life insurance, customers get rewards after a fixed time period or after a mishappening. In the above case study, various aspects or factors which influence the customers in purchasing the life insurance product have been covered. The above analysis can be amplified as per the national scenario with some specific modifications. This technique can also be extended to larger groups of customers. The choices of the customers may vary according to the conditions and influencing features. On the basis of above calculation, it is concluded that both the IF measures provide the same results for a decision-making problem. This method may be applied in other studies.

The idea of IF sets has immense significance across domains for decision-making and has better control over problems with uncertain conditions. It may be applied in other domains of study to achieve better results.

References

Adlassnig, K. P. (1980) A Fuzzy Logical Model of Computer-Assisted Medical Diagnosis". *Methods of Information in Medicine*, Vol. 19 (3), 141-148.

Adlassnig, K. P. (1982). "A Survey on Medical Diagnosis and Fuzzy Subsets". Eds.: Gupta M. M and Sanchez E., *Approximate Reasoning in Decision Analysis*, North-Holland Publishing Company, New York, Vol. 205, 203-217.

Ahmed, Y., Husain, S., and Alam, A. (2010). "Applying Intuitionistic Fuzzy Approach to Reduce Search Domain in an Accidental Case". *International Journal of Advanced Computer Science and Applications*, Vol. 1(4), 69-72.

Atanassov, K. T. (1986). "Intuitionistic Fuzzy Sets". *Fuzzy Sets and Systems*, Vol. 20(1), 87-96.

Atanassov, K. T. (1994). "Operators on Intuitionistic Fuzzy Sets". *Fuzzy Sets and Systems*, Vol. 64, 159-174.

Atanassov, K. T. (1994a). "New Operations Defined over Intuitionistic Fuzzy Sets". *Fuzzy Sets and Systems*, Vol. 61, 137-142.

Atanassov, K. T. (1999). "*Intuitionistic Fuzzy Sets" (Theory and Applications)*. Springer-Verlag Berlin.

Atanassov, K. T. (2012). "*On Intuitionistic Fuzzy Set Theory*". Springer-Verlag London.

Barik, B., and Patra, R. (2014). "Emerging Trends in Insurance-A Study in Indian Life Insurance Industry". *Abhinav National Monthly Refereed Journal of Research in Commerce and Management*, Vol. 3(6), pp. 7-12.

Bhandari, D., and Pal, N. R. (1993). "Some New Information Measures for Fuzzy Sets". *Information Sciences*, Vol. 67, 209-228.

Burillo, P., and Bustince, H. (1996). "Entropy on Intuitionistic Fuzzy Sets and on Interval-Valued Fuzzy Sets". *Fuzzy Sets and Systems*, Vol. 78(3), 305 -316.

Burillo P. and Bustince H. (1996a). "Construction Theorems for Intuitionistic Fuzzy Sets". *Fuzzy sets and systems*, Vol. 84, 271-281.

Chen, S. M. (1995). "Measures of Similarity between Vague Sets". *Fuzzy Sets and Systems*, Vol. 74, 217-223.

Cornelis, C. , Deschrijver, G., and Kerre, E. E. (2004). "Implication in Intuitionistic Fuzzy and Interval-Valued Fuzzy Set Theory: Construction, Classification, Application". *International Journal of Approximate Reasoning*, Vol. 35, 55-95.

Dash, G., and Sood, T. (2013). "Why Should One Invest in a Life Insurance Product-An Empirical Research". *Journal of Arts, Science and Commerce*, Vol. 4(1), 36-45.

De Luca, A., and Termini, S. (1972). "A Definition of a Non Probabilistic Entropy in the Setting of Fuzzy Sets Theory". *Information and Control*, Vol. 20(4), 301-312.

Deshmukh, K. C., Khot, P. G., and Nikhil. (2011). "Generalized Measures of Fuzzy Entropy and Their Properties". *International Journal of Engineering and Natural Sciences*, Vol. 10(3), 124-128.

Ejegwa, P. A., Akubo, A. J., and Joshua, O. M. (2014). "Intutionistic Fuzzy Set and its Application in Career Determination via Normalised Euclidean Distance Method". *European Scientific Journal*, Vol. 10(15).

Ejegwa, P. A., Akowe, S. O., Otene, P. M., and Ikyule, J. M. (2014). "An Overview on Intuitionistic Fuzzy Sets". *International Journal of Scientific & Technology Research*, Vol. 3(3), 142-145.

Granstrom, J. G. (2011). "*Treatise on Intuitionistic Type Theory*". Springer, Dordrecht, Heidelberg.

Hooda, D. S., and Jain, D. (2009). "Sub Additive Measures of Fuzzy Information". *Journal of Reliability and Statistical Studies*, Vol. 2(2), 39-52.

Horn, R. G. (1971). "Life Insurance Earnings and the Release from Risk Policy Release System". *Transactions of Society of Actuaties*, Vol. 23, 391-418.

Hung, W. L., and Yang, M. S. (2006). "Fuzzy Entropy on Intuitionistic Fuzzy Sets". *International Journal of Intelligent Systems*, Vol. 21(4), 443-451.

Husain, S., Ahmad, Y., and Alam, M. A. (2012). "A Study on the Role of Intuitionistic Fuzzy Set in Decision-making Problems". *International Journal of Computer Applications*, Vol. 48(19), 35-41.

Jha, P., and Verma, M. (2012). "Some New Measures of Intuitionistic Fuzzy Entropy and Directed Divergence". *International Journal of Fuzzy Mathematics and Systems*, Vol. 2(2), 433-439.

Kapur, J. N. (1997). "*Measures of Information and their Applications*". Wiley Eastern, New Delhi.

Kothari, H., and Agrawal, R. (2013). "A Study on Factors Individuals Investments towards Life Insurance Policies". *Altius Journal of Management, Commerce and Law*, Vol. 1(1), 51-61.

Kumar, A., Mahajan, S., and Kumar, R. (2011). "Two New Generalized Measures of Fuzzy Divergence". *International Journal of Mathematical Sciences and Applications*, Vol. 1(2), 831-838.

Kumar, V. (2011). "Application of Generalized Kapur's Entropy and Intuitionistic Fuzzy Sets (IFS) in Communication Systems and Medical Diagnosis of Headache". Ph. D. Thesis, University of Maharshi Dayanand, India.

Kumar, V. (2013). "Fuzzy Procedure for the Selection of Car Among Various Brands". *International Journal of Engineering Research and Technology*, Vol. 6(3), 337-342.

Liu, H. W. (2005). "New Similarity Measures between Intuitionistic Fuzzy Sets and between Elements". *Mathematical and Computer Modelling*, Vol. 42, 61-70.

Maji, P. K., and Roy, A. R. (2002). "An Application of Soft Sets in a Decision-making Problem". *Computers and Mathematics with Applications*, Vol. 44, 1077-1083.

Mints, G. (2002). "*A Short Introduction to Intuitionistic Logic*". Kluwer Academic Publishers, NewYork.

Neog, T. J., and Sut, D. K. (2011). "An Application of Fuzzy Soft Sets in Decision-making Problems using Fuzzy Soft Matrices". *International Journal of Mathematical Archive*, Vol. 2 (11), 2258-2263.

Prince, Gupta P., and Vijay, K. (2015). "Parametric Entropies in Intuitionistic Fuzzy Set Theory: A Hypothetical Case Study in Decision-making Problems". Mathematical Modeling, *Optimization and Information Theory*, Lambert Academic Publishers, 160-171.

Rényi, A. (1961). "On Measures of Entropy and Information". *Proceedings of the Fourth Berkeley Symposium on Mathematical Statistics and Probability*, Berkeley-Los Angeles, University of California Press. 547-561.

Ryjov, A. (2009). "The Measure of Uncertainty of Fuzzy Set's Collection: Definition, Properties and Applications". *IEEE Xplore*. 51-54.

Sanchez, E. (1977). "Solutions in Composite Fuzzy Relation Equation: An Application to Medical Diagnosis Brouwerian Logic". *Fuzzy Automata and Decision Process*, Elsevier.

Shannon, C. E. 1948. "The Mathematical Theory of Communication". *Bell's System Technical Journal*, Vol. 27, 379-423.

Sharma, V., Chowhan, S. S. (2013). "A comparative study of public and private life insurance". *Indian Journal of Applied Research.*, Vol. 3(1), 93-94.

Szmidt, E., and Kacprzyk, J. (2000). "Distances between Intuitionistic Fuzzy Sets". *Fuzzy Sets and Systems*, Vol. 114(3), 505-518.

Szmidt, E., and Kacprzyk, J. (2001). "Intuitionistic Fuzzy Sets in Some Medical Applications". In *5th International Conference on IFSs*, Sofia, Vol. 4, 58-64.

Szmidt, E., and Kacprzyk, J. (2001a). "Intuitionistic Fuzzy sets in Intelligent Data Analysis for Medical Diagnosis". *Proceeding of International Conference on Computational Science- Part II*, 263-271.

Szmidt, E., and Kacprzyk, J. (2001b). "Entropy for Intuitionistic Fuzzy Sets". *Fuzzy Sets and Systems*, Vol. 118, 467-477.

Szmidt, E. and Kacprzyk, J. (2003). "A Measure for Intuitionistic Fuzzy Sets". *Fuzzy Sets and Systems*, Vol. 121.

Szmidt, E. (2014). "*Distances and Similarities in Intuitionistic Fuzzy Sets*". Springer International Publishing, Switzerland.

Tsallis, C. (1988). "Possible Generalization of Bolzmann-Gibbs Statistics". *Journal of Statistical Physics*, Vol. 52, 479-487.

Upadhyay, P. (2013). "Satisfaction of the Policy Holders Protection in Insurance Sector". *International Journal of Advance Research in Computer Science and Software Engineering*, Vol. 3(2), 32-40.

Verma, R. K., and Sharma, B. D. (2011). "On Generalized Exponential Fuzzy Entropy". World Academy of Science, *Engineering and Technology*, Vol. 5, 12-24.

Verma, R. K., and Sharma, B. D. (2014). "Fuzzy Generalized Prioritized Weighted Average Operator and Its Application to Multiple Attribute Decision-making". *International Journal of Intelligent Systems*, Vol. 29, 26-49.

Wang, W., and Xin, X. (2005). "Distance Measures between Intuitionistic Fuzzy Sets". *Pattern Recognition Letters*, Vol. 26(1), 2053-2069.

Yadav, B., and Tiwari, A. (2012). "A Study on Factors Affecting Customers Investment towards Life Insurance Policies". *International Journal of Marketing Financial Services and Management Research*, Vol. 1(7), 106-123.

Zadeh, L. A. (1965). "Fuzzy Sets"; *Information and control*, Vol. 8, 338-353.

Zarandi, A. (2010). "On Fuzzy Information Theory"; *Indian Journal of Science and Technology*, Vol. 3(9), 1020-1025.

Zhao, H., Xu, Z. S., Ni, M. F., and Liu, S. S. (2010). "Generalized Aggregation Operators for Intuitionistic Fuzzy Sets". *International Journal of Intelligent Systems*, Vol. 25(1), 1-30.

Chapter 7

Information Entropy and Volatility Implied

Imen Mahmoud[1,*]
and Kamel Naoui[2]
[1]Laboratory of Research in Innovative Management Risk, Accounting and Finance (LARIMRAF), Department of Finance, University of Manouba UMA, Manouba Tunisia
[2]Dean of Graduate Business School of Tunis, Department of Finance, University of Manouba UMA, Manouba, Tunisia

Abstract

In this chapter, our aim is to test the efficiency of derivative markets by evaluating additional information about volatility implied in the Volatility Index, traded on the Chicago Board Option Exchange (CBOE). To this end, we used maximum entropy distribution which is a new method taken from statistical physics and introduced by Shannon (1948) as a model to assess risk-neutral density. Maximum entropy describes a belief in a competitive market - a market that consists of a large and a small number of rational investors while taking into account available information. Specifically, we propose a modified entropy version of Black and Sholes' pricing model (1973). Unlike Black and Scholes (1973), we do not hypothesize about the distribution of assets. Instead, we will use the maximum entropy distribution suggested by Buchen and Kelly (1996), Neri and Schneider (2012) to estimate random changes in asset prices. Our current study examines observations on the call option of the Volatility Index (VIX) traded on the Chicago Board Option Exchange (CBOE), over the period starting from March 14th, 2018. The obtained results led us to opt for the use of this new version of Black and Scholes' model (1973). Actually, this method allows for determining the

* Corresponding Author's Email: imen.mahmo@gmail.com

In: Understanding Information Entropy
Editor: Vijay Kumar
ISBN: 979-8-88697-839-1

arbitrary dynamics of equity and interest rates and volatility as well as implied volatility structures (Gulko, 1999, 2000, 2002).

Keywords: information, maximum entropy, probability distribution, volatility smile, option prices

1. Introduction

Options entropy theory is a generalization of Black and Scholes' options pricing model (1973) whose parametric hypotheses received lots of empirical criticism. For example, are the stock returns volatility always constant? Merton (1973) investigates Black and Sholes' assumptions and concludes that variance remains constant regardless of option contact lifespan and that the company's investment philosophy, capital structure, or dividend policy remain unchanged. Because, in this case, in Modigliani and Miller's old terminology (1958), the company changes its perception of "risk class," therefore, the dynamics of its stock price. Suppose the cash-flow of a company that does not change its investment policy, capital structure and dividend policy during the time interval (0, T). The price appears to be a fixed-income instrument, as it is not likely to fluctuate over the next few months (0,T). However, the price volatility of fixed income products such as bonds varies over time. As a result, the volatility of stock returns is not consistent. Furthermore, the current dynamics of stock returns may not be represented by a Brownian movement and a constant variance. Some authors believe that stock returns and volatility fluctuate at random. Hull (1987) and Goldenburgh (1988) are two examples of these authors (1991). At the expense of extra parameters, these authors suggested models that are more generic and presumably more accurate. Traditional volatility's time characteristics (i.e., the term-based structure of implied volatilities) are inconsistent with market data, according to Bates (2003).

2. Analytical Framework

2.1. Literature Review

To price derivatives, Mandelbrot (1962) and Samuelson (1965), Sharpe (1964), Black and Scholes (1973), Merton (1976) as well as Cox and Ross

(1979) used random processes represented in normal distribution. These processes seem to determine the most appropriate structure of markets to develop derivatives pricing theory (Black and Scholes, 1973).

Derivatives markets have known significant growth since the mid-80's. However, the sudden interest in risk-neutral densities may be explained by the series of crises triggered by the "subprime crisis." An important outcome of this is the concern about analyzing information contained in derivatives prices. Future prices can be considered as a price at a time in the future. Thus, such a pricing behavior implies anticipating prices in the future, bringing the concern with crises at the center of interest, in particular, the concern with estimating probabilities through assets distribution of corresponding options.

For a long time, the famous Black and Scholes derivatives pricing and hedging model (1973) has been considered as the unique pricing model. The BS model assumes that asset prices volatility is constant and independent from the attributes of an option maturity and the strike price. This model refers to the restricted hypothesis that RND of assets follows the log-normal distribution for the pricing and hedging of derivatives. Since the 1987 stock market crash, the index implied volatility for currency-based options prices has decreased as a function of the strike price. This phenomenon consists in using the strike price to trace volatility - an instrument known as volatility smile. Volatility smile is a bijection of the implied distribution of similar options. Such a phenomenon is fruitful when returns exhibit fat tails and their fluctuations correlate. This will most likely affect options pricing in the Black-Scholes model. Following the first tests on option prices, like those carried out by Robinstein (1985), the Black and Scholes model (1973) was strongly questioned. Much like in a downward market, the real hedge cost for a trader selling these "puts" was much larger than the premium received. This is likely to happen to asset prices when the model's parameters are difficult to observe or estimate in an accurate way. Several studies used a number of methods to determine options prices density and therefore explain how the BS model empirically failed (Garia et al. 2019). These methods can be classified into three categories: non-parametric models, structural models and non-parametric models. These models assume no hypothesis on assets distribution (Rubinstein, 1994; Ait- Sahalia and Lo, 1998; Madan and Milne, 1994). Indeed, structural models proceed through making a hypothesis about the functional form of distribution (Heston, 1993; Bates, 1996). Similarly, diffusion models proceed through a stochastic temporal description of underlying prices. In this study, we will opt for non-parametric models, namely the Maximum Entropy Density (MED). Applying entropy to study economic phenomena is relatively new and offers a variety of models to examine uncertainty. Shannon and Weaver (1949) state that "the exact quantity necessary to set 'information' corresponds exactly to the thermodynamic concept of entropy". Similarly, Dionisio et al (2001) state that "the use, as a measure, of uncertainty in finance seems to have lots of potential and a golden opportunity for both theoretical and empirical developments". Such a concern with this measure translates the importance of uncertainty in econophysics. Unlike economists who define and exclude change by a single model, econophysicists use this concept epistemically and study its different dimensions. Each entropy presents a specific statistical framework for modeling the reality under investigation. Economophysicists provide a range of operational methods for studying uncertainty by giving this diversity. Likewise, as far as finance is concerned, uncertainty can be approached through the use of statistical measures in physical and information theory, namely in Shannon Entropy (SE). Shannon Entropy (1949) is a mathematical function which represents the amount of information in a data set. It is a measure of the degree of uncertainty associated with a random occurrence or, more accurately, its distribution. In fact, the information produced by each new occurrence is a function of the event's uncertainty. Clausius (1865) was the first to use it in thermodynamics to estimate the ratio of heat transmitted during a reversible process in an isolated system, and later, around 1900, in statistical physics by Boltzmann and Gibbs. Statistically, interpreting entropy amounts to measuring the remaining uncertainty in a system after macroscopically observing its properties (pressure, temperature or volume). By the mid-twentieth century, it found its way in engineering and mathematics, Shannon's work in engineering and communication, as well as Kolmogorov's work in probability theory and dynamic systems theory, have all influenced me. The information conveyed from a source system is a stochastic cybernetic system, with the message acting as a random variable. The amount of information contained in a message is defined by entropy, which qualifies the expected value of the information contained in the message. However, information theory has used the concept of entropy to quantify the degree of uncertainty and disorder in a message's information. It is in this context that entropy was used in economic theory. Some researchers such as Gulko (1998) and Dionisio (2001) consider entropy as a measure of information because it shows the real value of information in financial markets. Borrowed from the widely used statistical physics to quantify disorder and uncertainty in dynamic systems, this approach has also been applied in financial theory and several other fields (market efficiency, asset pricing and derivatives portfolio Management).

Gulko (1999) was the first to suggest employing entropy to examine financial time series, demonstrating that the maximum entropy formalism, also known as informational efficiency, allows the market efficiency hypothesis to be tested and validated. Zhou et al. (2013) evaluated the relevant entropy concepts and principles that have been used in finance, in particular those related to derivatives pricing. The authors have shown that entropy has several unique advantages over measuring uncertainty and describing its distribution. The concepts and principles of entropy can also be used to measure derivatives, which would contribute to the general problem of determining the distribution probability of limited information and incomplete markets. A Bayesian statistical inference is a method of estimating density utilizing partial information in the form of a finite number of time series as constraints. Density of maximum entropy (MED) is a view consistent with information on prices, i.e., extracting the maximum without being affected by missing or unknown information (Jaynes, 1957). Considering skewness and kurtosis (higher order moments), in decision-making becomes necessary since distributions of financial assets are not normally distributed. Mandelbrot (1962, 1963), Fama (1963), Fama and Roll (1968) conclude that investors are sensitive to information in higher-order moments. As a result, without providing any underlying theoretical framework, this density estimate method can provide a plausible approximation of RND based on all the data available on option prices. This approach generates a strictly positive probability through a construction process, which is one of its key advantages over other RND recovery methods such as Edgeworth's series expansion and that of Corrado and Su (1996).

Compared with other existing risk neutral density recovery methods which use the maximum entropy principle, the advantage of the latter suggests that these moments can be recovered from options prices at each moment, first by using efficiently all available information in the market.

Accordingly, several authors have applied different versions of the maximum entropy principle to recover RND of options prices. Buchen and Kelly (1996) were among the first to use this concept to RND data on prices of European options at any time, by imposing constraints that prove that the observed option prices are expectations of future earnings.

They proposed this version of maximum entropy to derive the probability distribution of a call price. Stutzer (1996), on the other hand, used asset observations to anticipate the probability distribution of asset returns. After that, the author employed the greatest entropy concept to calculate RND while keeping in mind that the asset is a martingale. Rompolis (2010) proposes a new method to use the maximum entropy principle to recover the RND of options prices. The main feature that distinguishes Rompolis from those mentioned above is the way information in the option price has been modeled. This solution solves the MED problem by using non-central moments of inferred RND and a cross-section of European options prices as constraints, rather than relying on option prices values. To recover the RND of options prices, Neri and Schneider (2012) applied the greatest entropy principle. They compared their results with those of Buchen and Kelly (1996). The estimated density is, in turn, used to interpolate implied volatilities. They showed that this approach meant a very realistic volatility layer for options on the S & P 500 Index in the money.

Bartiromo (2013) used this approach for price returns to extract implied volatility of an option price of a share. This distribution relates to an entropic interpretation of the arbitration absence principle (Gulco, 1999) and gives a sufficiently good measure of the probability of significant fluctuations. Therefore, it should be preferred to the Gaussian distribution for financial derivatives pricing. In this chapter, we use the MED technique to study asset pricing process and its contribution regarding information content. This method, borrowed from information theory, has been widely used in theory. Once density is constructed, it will be used to study volatility. Specifically, we focus on the price distribution of a given asset at a future time via the maximum entropy distribution (MED) of assets. Density is obtained for the intervals between the strike prices. However, unlike Black-Scholes model, we make no hypothesis on the distribution of assets. Instead, we maximize Shannon's Boltzmann entropy to derive a distribution that is compatible with the option pricing procedure. The goal of this maximizing problem is to find a collection of Lagrange multipliers. Density can then be used to interpolate implied volatilities, and a volatility layer can be created by repeating this process for a variety of maturities. The results match the volatility layers seen in the markets surprisingly well. Furthermore, an analytical formula for the price of an asset at a certain period is developed. We can recover the smile of implied volatility for the call of the Volatility Index (VIX), negotiated on the Chicago Board Options Exchange, by calculating numerous strike prices for different maturities (CBOE)[1].

[1] Chicago Board Options Exchange Website (www.cboe.com): CBOE Binary Option.

2.2. Method

2.2.1. Sample Construction

In this study, we use, as a first leg, March 14th, 2018 for the call option VIX (Volatility Index) traded on the Chicago Boards Options Exchange. Our data is taken from www.cboe.com.

2.2.1.1. Estimating the Maximum Entropy Density (MED)

The probability distribution obtained is, then, used as *a priori* probability, in a conventional Bayesian inference process.

If the discrete random variable X's distribution is as follows:

$$H(X) = -\sum_{i\geq 1} p_i \log p_i \text{ and i=1,2,....} \quad (1)$$

If X is a continuous random variable, with a probability density p(x), then entropy is equal to:

$$H(p) = -\int_0^\infty p(x)\ln p(x)dx \quad (2)$$

Buchen and Kelly (1996) propose a similar method to study a density P (x) that maximizes entropy under constraints for European options prices, a pay-off for different strikes $K_0 = 0, K_1, K_2, ..., K_n$.

Let X be a discrete random variable that represents strictly increasing strikes $K_0 = 0, K_1, K_2, ..., K_n$, at a fixed maturity T.

Distribution entropy is given by the following equation (2):

With the constraints of payoff for a call is

$$E(S_T - K_i) = C_i \text{ , } \int_{K_i}^{\infty} (x - K_i)p(x)dx = C_i \quad (3)$$

for P(X) to be a density, we should impose $p(x) \geq 0$

$$\int_0^\infty p(x)dx = 1 \quad (4)$$

Buchen and Kelly (1996) impose two additional arbitration constraints. On the one hand, they require that the expected discounted price is equal to the observed price. Moreover, the expectation of the underlying price equals the forward rate at *Time_t*. These constraints can be written as follows

$$E(S_T - K_i) = C_i \ , E\left[C(x)\right] = \int_0^\infty p(x)C_i(x)dx = d_i \tag{5}$$

Is satisfied since d: the forward price of S for time T, and the functions *C (x)* are given by:

$$\begin{aligned} &C_0(x) = x \\ &C_i(x) = e^{-rT}(x - K_i) \end{aligned} \tag{6}$$

In particular, these two constraints for i = 0 mean that p(x) is the density, and from the second constraint it immediately follows that:

$$E(S_T) = \int_0^\infty xp(x)dx = C_0 \tag{7}$$

Observing a call with a strike price K_i , K_{i+1} , high at K_i the level and S_T is between K_i , and K_{i+1} zero otherwise, we get the following condition:

$$-\int_0^\infty p(x)\ln p(x)dx = \sum_{i=0}^{n}(-\int_{K_i}^{K_{i+1}} p(x)\ln p(x)dx \tag{8}$$

We formally derived from the phrase density which maximizes the entropy. This maximization problem reports to looking for a set of Lagrange multipliers on the interval $\left[K_i, K_{i+1}\right]$:

$$\varphi(p) = -\int_{k_i}^{k_{i+1}} p(x)\ln p(x)dx \tag{9}$$

Under the constraints

$$\vartheta(p) = \int_{k_i}^{k_{i+1}} p(x)dx \tag{10}$$

$$\theta(p) = \int_{K_i}^{K_{i+1}} xp(x)dx \tag{11}$$

The number of the to-be-maximized constrained optimization problems assuming an entropy measure is given in Equation (2).

This is to solve the following equation:

$$\delta\varphi(p) + \lambda_1 \delta\vartheta(p) + \lambda_2 \delta\theta(p) = 0 \tag{12}$$

where $\lambda, \lambda_1, \ldots.., \lambda_N$ are the to- be determined Lagrange parameters.

Therefore, in the interval, we should have

$$\ln p(x) + 1 = \lambda_1 + \lambda_2 \tag{13}$$

By introducing $\alpha = e^{\lambda - 1_1}$ and $\beta = \lambda_2$, we obtain the value of the maximum entropy distribution, given by:

$$p(x) = \alpha e^{\beta x} \tag{14}$$

Like other researchers, we formally derived from the above function the density that maximizes entropy using the Lagrange multipliers method.

In a call option with a strike price K1 = 0, i.e., of all the data, only the distribution mean is given. Since the total mass is equal to 1 which is also known, we have two primary constraints necessary to enforce our arguments, and we can, therefore, maximize the entropy distribution of a given option.

Furthermore, we set a constant volatility at 25%. For each maturity T> 0, we calculate the price using Black-Scholes' model. Then, we apply the approach of maximum entropy for different strike prices. In other words, we find the density maximizing the entropy for 1, 3 and 5 strike prices.

3. Results and Discussion

3.1. Descriptive Statistics

The knowledge of higher order moments (variance, skewness and flattening kurtosis allows us to deduce the likely market price. The skewness indicates the direction of anticipated movements. This coefficient is different from zero for calls calculated using the Black-Scholes' model and the entropy model. This reflects an asymmetry. This asymmetry may be visible when plotting the implied volatility based on strike price if the smile has a more elongated leg than the other. In this case, we deduce that traders anticipate lower prices rather than a rise. The Kurtosis coefficient is less than 3: here, we are talking about a platykurtic distribution; and this reveals the possibility of slight changes up to the maturity date of the option. Similarly, the skewness coefficient (the effect of kurtosis) appears in the phenomenon of Smile. Thus, investors anticipate a slight decrease rather than a decline.

Table 3 reports the options prices under study and density-describing parameters. In practice, however, implied volatilities are not flat, as shown in the example above. They differ from strike prices and maturities. Such a finding is discussed in detail in Gatheral (2006).

This test is used as a non-parametric alternative to the t-Student test of two dependent samples. Specifically, for two independent samples if the hypotheses of the t-student test are not valid, i.e., if the variable is normally distributed and if equal variance is not met, then the Wilcoxon test is recommended. The Wilcoxon test is a non-parametric test that compares two series, with the null hypothesis of equality of means.

Referring to the results, we found no significant difference between the pricing call options by Black and Scholes 'model, using the proper distribution of the asset, and the pricing call options by the maximum entropy distribution, concerning the strike prices 1, 3 and 5.

Figure 1 represents the three densities 1, 3 and 5 strikes, obtained by maximizing entropy distribution. Importantly, with five strike prices, the adjustment of the distribution is better.

Table 1. Descriptive statistics

	Median	Maximum	Minimum	Standard deviation	Skewness	Kurtosis	Jarque-Bera	Probability	Observations	
Call_B &S	9.887362	3.481679	35.00000	0.010251	12.16446	0.917697	2.387865	2.339613	0.310427	15
Call with MED (1 Stike)	9.930485	3.481679	35.00000	0.037674	12.13043	0.921130	2.394028	2.350702	0.308711	15
Call with MED (3 Stikes)	9.928480	3.481679	35.00000	0.037674	12.13062	0.921538	2.394507	2.352217	0.308477	15
Call with MED (5 Stikes)	10.27133	4.348056	35.00000	0.212268	11.88625	0.934389	2.436943	2.380853	0.304091	15

Table 2. Option prices and density parameter

K	**0**	**5**	**10**	**15**	**20**	**25**	**30**	**35**	**40**	**45**	**50**	**55**	**60**	**65**	**70**
Call B&S	35,0000	30,0000	25,0000	20,0005	15,0288	10,3035	6,3243	3,4817	1,7465	0,8143	0,3595	0,1526	0,0630	0,0256	0,0103
1 Strike															
alpha	4,3561E-06					0,6629									
Beta	0,3656					-0,0863									
Call with MED	35,0000	30,0001	25,0011	20,0076	15,0486	10,3035	6,6933	4,3481	2,8246	1,8349	1,1920	0,7743	0,5030	0,3268	0,2123
Implied volatility		0,5116	0,3860	0,3142	0,2693	0,2500	0,2840	0,3127	0,3313	0,3439	0,3528	0,3591	0,3637	0,3671	0,3695
K	0	5	10	15	20	25	30	35	40	45	50	55	60	65	70
Call B&S	35,0000	30,0000	25,0000	20,0005	15,0288	10,3035	6,3243	3,4817	1,7465	0,8143	0,3595	0,1526	0,0630	0,0256	0,0103
3 Strikes															
alpha	6,4755E-11			1,1944E05		0,0155		5,3814							
beta	1,0679			0,3216		0,0345		-0,1293							
Call with (MED)	35,0000	30,0000	25,0000	20,0005	15,0375	10,3035	6,3456	3,4817	1,8238	0,9553	0,5004	0,2621	0,1373	0,0719	0,0377
Implied volatility		-1,0000	0,2713	0,2500	0,2593	0,2500	0,2520	0,2500	0,2560	0,2645	0,2717	0,2774	0,2819	0,2853	0,2880
K	0	5	10	15	20	25	30	35	40	45	50	55	60	65	70
Call B&S	35,0000	30,0000	25,0000	20,0005	15,0288	10,3035	6,3243	3,4817	1,7465	0,8143	0,3595	0,1526	0,0630	0,0256	0,0103
5 Strikes															
Alpha	6,4755E-11			2,9663E07	0,0001	0,0038	0,0632	5,3814							
beta	1,0679			0,5180	0,2419	0,0856	-0,0087	-0,1293							
Call with (MED)	35,0000	30,0000	25,0000	20,0005	15,0288	10,3035	6,3243	3,4817	1,8238	0,9553	0,5004	0,2621	0,1373	0,0719	0,0377
Implied volatility		-1,0000	0,2713	0,2500	0,2500	0,2500	0,2500	0,2500	0,2560	0,2645	0,2717	0,2774	0,2819	0,2853	0,2880

Table 3. Distribution parameters

Number of Strikes	Mass	Mean	Entropy
1	1	35	3,6384
3	1	35	3,5645
5	1	35	3,5623

Table 4. Wilcoxon Test

		Observations	Mean	Std. Err.	Std. Dev.	[95% Conf. Interval]		Probability
Call with MED	1 Strike	15	0.3839701	0.1042918	0.4039206	0.1602863	0.60765	0.0012
	3 Strikes	15	0.0431229	0.0135861	0.0526188	0.0139836	0.0722622	0.0034
	5 Strikes	15	0.0411183	0.0139325	0.0539603	0.0112361	0.0710006	0.0053

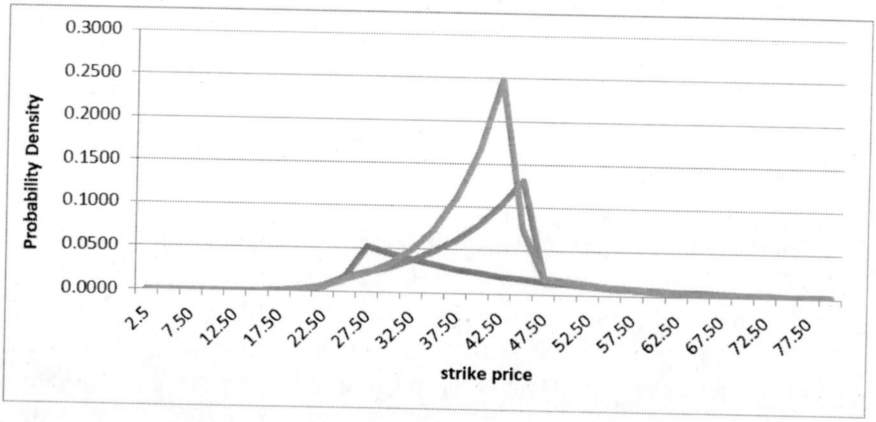

Figure 1. MED for March 14th (VIX).

The results are not surprising, given the density modeling method used. We clearly see that the exponential parts of densities are equal to the number of observed prices. It seems that this technique admits the presence of a large asymmetry, but reveals a kurtosis excess. Our results are consistent with those of Buchen and Kelly (1996), yet, they slightly differ from those of Neri and Schneider (2012).

Eliminating parameters while maximizing entropy suggests that many processes are leading to the same maximum entropy probability density at time T. Therefore, maximum entropy density is relatively robust. MED formulae readily lend themselves to empirical tests. Their structure and simplicity are similar to those of Black and Scholes' assets pricing model. They also take many properties after this model as they check the put-call parity (Neri and Schneider, 2012).

After finding densities by maximizing entropy, we used them to interpolate implied volatilities, by repeating the process over a range of maturities. Now, by varying t, we can construct a volatility layer which is, again, similar to that observed in the stock markets. MED and the lognormal density are qualitatively similar. Both of them converge to zero and they both have a bell shape. Bell-shaped densities adequately describe random values. It is well-known that the most important anomaly, in option prices, is volatility smile. This stylized fact shows the shape of a smile when plotting implied volatility, as a function of the strike price calculated from the options prices observed on the market. Volatility "Smile" indicates, on the one hand, that asset returns have skewed distributions or fat tails (by opposition to the normality hypothesis of securities prices) and that, on the other hand, fluctuations of these returns correlate. According to the Black and Scholes' model, we should get a straight horizontal line that implies that all call options of the same stock with the same expiry date but with different strike prices, should have the same implied volatility. Obviously, this is not the case in real options markets.

Thus, implied volatility, concerning real options prices, under the Black and Scholes 'model, is not constant; it decreases with the strike price K. Our results depict well the volatility layers observed on the market. We can, therefore, recover the volatility smile. Against the above, it can be said that the MED option pricing model can replace the lognormal distribution. Such a finding suggests that MED can be used for term- contract options pricing of commodities. Black's model assumes that term-contracts on commodities follow a geometric Brownian movement while the short-term interest rate is constant. MED does not impose such restrictions. The "Smile" takes, then, a curved shape. In fact, market participants are willing to pay more for one option. Then, at the first maturity, the lower the strike price is, the higher the volatility is. Investors expect a decrease in prices rather than a rise, in the short-term. In the chart, the U-shape is not the right curve. They anticipate a slight price drop rather than a *sudden low* which further explains skewness. The maximum entropy principle to estimate the probability distribution of an asset may well be accepted as a model for price options.

Conclusion

The information contained in option prices have received the attention of financial institutions and central banks. The aim is to consider market reactions against fixing interest rates. Investors may, then, understand market expectations about future prices and use them to manage their portfolios.

In this chapter, we tried to test the entropy market hypothesis (1999) to understand the participants ‘market behavior. To this end, we presented a modified version of maximum entropy in order to price options in both complete and incomplete markets. The B and S option pricing model is only valid for full and perfect markets. When the market is incomplete or imperfect or even both, the no-arbitration principle is adopted, by applying restrictions on open prices that are too low to be useful. First, we tried to model density by maximizing entropy for VIX option. Once this density is obtained, it was then used to get a volatility layer. Entropy can be considered as an unbiased measure of the probability distribution, thus maximizing entropy. Our proposal consists in finding an equal probability distribution that is consistent with market information. Implied volatility is the on parameter in the model which may be obtained by analyzing options prices. Our contribution, then, aims at proposing an analytical expression to estimate probability density function for prices that can be used to extract the value of implied volatility of call options. This distribution ultimately relates to interpreting the entropy principle of no arbitration and gives a sufficiently good estimate of the probability of significant fluctuations. Therefore, in our opinion, it is preferable to a Gaussian distribution when it comes to pricing derivatives. The primary objective of modern finance is assets. Choosing a pricing model depends on available information. Entropy is one of the central concepts of information theory, and market participants react to information when taking decisions. To model such reactions, we believe that entropy is a natural tool for that aim. The method we proposed can be used reliably and efficiently in practice. Most importantly, the process produces a remarkably realistic volatility layer, unlike the Black-Scholes’s model revealing a steep incline for short maturities which, in turn, decay with longer maturity. Therefore, this entropy-based distribution is to be preferred to the Gaussian distribution when pricing financial derivatives are concerned. The approach presented, in this study, is particularly useful to analyze derivatives as it does not consider the entire market to determine risk-neutral probabilities. Indeed, our chapter contributes, significantly, in clarifying the “volatility smile” puzzle and, therefore the asset pricing process.

References

Bartiromo, R. (2013). Maximum entropy distribution of stock price fluctuations. *Physica A*, 392(7), 1638–1647.

Bates, D. (1996a). Testing Option Pricing Models, in Handbook of Statistics. *Statistical Methods in Finance*, (14) 567-611.

Bates, D. (2003). Empirical Option Pricing: A Retrospection. *Journal of Econometrics,* 116, 387-404.

Black, F. and Sholes, M. (1973). The pricing of options and corporate liabilities. *Journal of political economy*, 81(3), 637-654.

Borwein, J., Choksi, R., and Marechal, P. (2003). Probability Distributions of Assets Inferred from Option Prices via the Principle of Maximum Entropy. S*IAM Journal on Optimization*, 14(2), 464–478.

Brody, D. C., Buckley, I. R. C. Constantinou, I. and Meister, B. (2005). Entropic Calibration Revisite. *Physics Letters A*, 337, 257–264.

Buchen, P. W. and Kelly, M. (1996). The Maximum Entropy Distribution of an Asset Inferred from Option Prices. *Journal of Financial and Quantitative Analysis* ,31(1), 143–159.

Chalamandaris, G. and Romplis, L. S. (2012). Exploring the role of the realized return distribution in the formation of the implied volatility smile. *Journal of banking and finance*, 36(4), 1028-104.

Clausius, R. (1870). On a mechanical theorem applicable to heat. *The London, Edinburgh, and Dublin Philosophical Magazine and Journal of Science*, (40), 122–127.

Cox, J. C., Ross, S. A., and Rubinstein, M. (1979). Option pricing: A simplified approach. *Journal of Financial Economics*, (7) 145–166.

Dionísio, Adreia. 2001." *Entropy Analysis as Uncertainty Measure and Information Value in Portuguese Stock Market*." Master Thesis, Evora University.

Goldenberg, D. H. (1991). A unified method for pricing options on the diffusion process. *Journal of Financial Economics*, 29, 3–34.

Gulko, L. (1999). The Entropic Market Hypothesis. *International Journal of Theoretical and Applied Finance*, 2(3), 293–329.

Gulko, L. (2000). Information and entropy in incomplete markets. *International Journal of Theoretical and Applied Finance*, 3(3), 561.

Gulko, L. (2002). The Entropy Theory of Bond Option Pricing. *International Journal of Theoretical and Applied Finance,* 5(4), 355–383.

Heston, S. L. (1993). A Closed-form solution for options on assets with stochastic volatility with application to bond and currency option. *Review of financial studies*, 6(2), 327-343.

Hull, J. C. and White, A. D. (1987). The pricing of options on assets with stochastic volatilities. *Journal of Finance*, 42(2), 281-300.

Ingersoll, J. E. 1987. Theory of Financial Decision-making, Rowman & Littlewood Publ. Inc. *Savage, Maryland*, 496.

Jaynes, E. T. (1957). Information Theory and Statistical Mechanics. *Physics Reviews,* 106, 620–630.

Mandelbrot, B. (1963). The variation of some other speculative prices. *Journal of Business*, 36, 394-419.

Merton, R. C. (1973). Theory of rational option pricing. *Bell J. Economics and Management Science*, 4, 141–183.

Modigliani, F. and Miller, M. (1958). The cost of capital, corporation finance and the theory of investment, The Amer. Economic Rev, 84, 261-297.

Neri, C. and Schneider, L. (2012): Maximum entropy distributions inferred from option portfolios on an asset. *Finance Stochastics*, 16, 293–318.

Rompolis, L. S. (2013). Retrieving risk neutral densities from European option prices based on the principle of maximum entropy. *Journal of Empirical Finance*, 17(5), 918–937.

Ross, S. A. (1977). Risk-Return, and Arbitrage. In: Friend, I. and J. Bicksler (Eds.). *Risk and Return in Finance I*, Ballinger, Cambridge, M.A.

Rubinsein, M. (1985). Nonparametric tests of alternative option pricing models using all reported trades and quotes on the 30 most active CBOE option classes from August 23, 1976, through August 31, 1978. *Journal of Finance*, 40(2), 455-480.

Samuelson, P. A. (1965). Proof that properly anticipated prices fluctuate randomly. *Industrial Management Review,* 6, 13–31.

Shannon, C. E. (1948). A Mathematical Theory of Communication. *System Technical Journal*, 27, 379–423, 623–656.

Sharpe, W. F. (1964). Capital asset prices: A theory of market equilibrium under conditions of risk. *Journal of Finance,* 19 (3), 425-442.

Stutzer, M. (1996). A simple nonparametric approach to derivative security valuation. *Journal of Finance*, 51, 1633–1652.

Zhou, R., Cai, R. and Ton, G. (2013). Applications of Entropy in Finance. *A Review Entropy*,15, 4909-4931.

Index

L

M

N

O

P

Q

R

S

T

U

V

W

About the Editor

Dr. Vijay Kumar received his PhD in Statistics from Maharishi Dayanand University, Rohtak. Presently, he is working as a Professor and Head, Department of Applied Sciences, Manav Rachna International Institute of Research & Studies, Faridabad Haryana, India. His research interest includes Fuzzy information theory, Decision-making and Machine Learning. He has authored more than 65 research publications including research papers, conference papers and chapters in Books. He has authored 3 books, out of which 2 are research books. He gave several invited talks in the conferences of national and international repute. Prof. Vijay Kumar serves as editorial board member for several International/National journals and reviewer for many National and International journals.
Email: drvijaykumarsudan@gmail.com

About the Contributors

Chapter 1

Amlan K. Roy is working as Professor with Indian Institute of Science Education and Research, Kolkata. His research interest includes Electronic structure theory, Density functional theory, Information theory in chemistry, Confined quantum systems. He has published more than 50 research papers in national and international journals.

Sangita Majumdar is research associate with Indian Institute of Science Education and Research, Kolkata, India

Neetik Mukherjee is Postdoctoral Fellow at Indian Institute of Science Education and Research (IISER), Kolkata, India.

Chapter 2

Denis Sabirov got his M.Sc. at the Magnitogorsk State Technical University (2006) and received his PhD in 'physical chemistry' and 'mathematical and quantum chemistry' from Bashkir State University. Now he is the head of the Laboratory of Mathematical Chemistry of this institution (since 2018). His research involves the prediction of molecular properties and reactivity by means of quantum chemical methods and focuses mainly on fullerenes, polycyclic aromatic hydrocarbons and their derivatives. He tries to incorporate the computational results into the relevant studies in the fields of experimental chemistry, materials science and astrochemistry.

Chapter 3

Pratiksha Tiwari is currently working as Sr. Assistant Professor in the Department of Management at Delhi Institute of Advanced Studies. She is NET qualified and holds PhD degree in Statistics from Maharshi Dayanand University, Rohtak, Haryana, India. Her areas of specialization in teaching and research are Fuzzy information theory, Quantitative Methods, Operations Research and Business Research. She has 20 years of teaching experience and coordinating the Placement Committee at DIAS. She has published papers in various referred SCOPUS indexed journals. She has also completed a minor research project sponsored by Indian Council of Social Science Research (ICSSR) and working on another is research project sponsored by IMPRESS.

Chapter 4

H. D. Arora is serving as a Professor and Formerly Head of Mathematics Department, Amity Institute of Applied Sciences, Amity University, Noida, India. An M. Phil in Operations Research, M.B.A. in Marketing, PhD in Information theory, Dr. Arora is a Gold Medalist in M. Sc. Operations Research. He has a vast experience of more than 22 years in the education industry. He has published more than 12 books in the field of Engineering Mathematics, Business Statistics, Modelling and Computing, and allied areas. Dr. Arora has made valuable contribution to research in Fuzzy information theory, Computational Mathematics, Software Reliability and guided more than 7 students for PhD program. He has several publications to his credit in National and International Journals of high repute. Presently he is working in the field of Fuzzy Information theory, Computational Mathematics, Coding Theory and Hardware Reliability etc.

Anjali Naithani is working with Department of Applied Mathematics, Amity Institute of Applied Sciences, Amity University, Noida, India. She has more than 20 publications and is a member of many professional bodies.

Rajender Kumar is presently working as Associate Professor in the Department of Mechanical Engineering, FET, MRIIRS. He is actively involved and associated with various national level societies promoting Engineering/Technical/Quality concepts in India such as IEI (Member Local Chapter Committee), IIIE (Senior Member), QCFI (Governing Council

Member), QCI (Registered Lead-Auditor), IIF (Member and Former Honorary Secretary, Faridabad Chapter), ISHRAE etc. He has received a grant from Government of India to conduct activities related to Entrepreneurship Development. He has 55 research papers, four book chapters and two published patents in his credit. For his contribution, he is a recipient of the Young Engineer-2021 and Young Personality of the Year-2022 awards. His area of expertise is Industrial Engineering and Management, Entrepreneurship and Business Incubation, SCM, Lean, Green and Agile Manufacturing System.

Chapter 5

Vijay Kumar is working as a Professor and Head, Department of Applied Sciences, Manav Rachna International Institute of Research & Studies, Faridabad Haryana INDIA. His research interest includes Fuzzy information theory, Decision-making and Machine Learning. He has authored more than 65 research publications and has authored 3 books.

Sarika Jain is working as an PGT teacher with Education department, Government of Haryana India. She is well published and a involved in research in decision-making techniques.

Chapter 6

Prince Goyal is working as an Assistant Professor in the department of Community Medicine, Dr. Yashwant Singh Parmar Govt. Medical College, Nahan, Himachal Pradesh India. He received his Ph.D in Statistics from Maharshi Dayanand University, Rohtak India. He has published more than 17 researcher papers and books of national and international repute and is a life member of various societies.

Chapter 7

Imen Mahmoud is a researcher at Laboratory of Research in Innovative and Management, Risk, Accounting and Finance, University of Mannouba. Dr Imen Mahmoud received her PhD in Finance from University of Tunis El

Manar. She published a number of papers in preferred journals and participated in a range of forums on Finance. She also presented various academic as well as research-based papers at several national and international conferences. In addition to teaching in the School of Business on portfolio management, derivatives and capital markets. Dr. Imen Mahmoud is focused on research around the physical foundation of economics and social sciences. Dr. Imen Mahmoud keeps improving his "analytical thermodynamic theory" and uses it in a variety of contexts, primarily in the realm of finance

Kamel Naoui is a researcher at Laboratory of Research in Innovative and Management, Risk, Accounting and Finance, University of Mannouba. He received his PhD in Finance from University of Tunis El Manar. His research interests include Risk Management, Corporate finance, International finance, Capital markets and institutions, Banking and market microstructure, Islamic Finance and Behavior Finance.